Current Practices and Trends in Technical and Professional Communication

About the ISTC

With a history dating back more than 70 years, the Institute of Scientific and Technical Communicators (ISTC) is the largest association representing information development professionals in the UK. It:

- Encourages professional education and supports standards
- Provides guidance about the value of using professional communicators
- Facilitates research, discussion and networking opportunities
- Liaises with other national and international technical communication associations.

The Institute's members create information that has an impact on people in virtually every sector of industry and society. They develop a wide range of information products, such as installation instructions for engineering equipment, operating and safety procedures for consumer products, and online user guides for computer software.

In their work, they face varied challenges that have an impact on their responsibilities, day-to-day tasks and careers. The ISTC provides opportunities for them to understand and overcome these challenges while fulfilling their core responsibility of communicating complex or important information in the most effective way for their readers.

One of the ways in which the Institute achieves this is to publish information in its monthly newsletter, *InfoPlus*, and its award-winning quarterly journal, *Communicator*. It is also the publisher of a series of books, covering topics that have specific relevance to those working in technical communication.

If you would like to write for this series, please send your proposal to books@istc.org.uk.

For more information about the ISTC, visit www.istc.org.uk.

Current Practices and Trends in Technical and Professional Communication

Edited by Stephen Crabbe

Institute of Scientific and Technical Communicators

Current Practices and Trends in Technical and Professional Communication

First published in Great Britain in 2017 by the Institute of Scientific and Technical Communicators (ISTC).

ISBN-10 09506459-90

ISBN-13 978-0-9506459-9-5

Reprinted 2019

A CIP record for this book is available from the British Library.

Typographical arrangement © Institute of Scientific and Technical Communicators, 2017

Typeset in Lucida Bright, Myriad Pro and Consolas by Alison Peck

Indexed by Katherine Judge

Cover adapted by Alison Peck from an original design by Matthew Jennings

Institute of Scientific and Technical Communicators (ISTC) Unit 19, Omega Business Village, Thurston Road, Northallerton, DL6 2NJ, United Kingdom

Telephone: +44 (0) 20 8253 4506

E-mail: **istc@istc.org.uk**

Visit our website: **www.istc.org.uk**

Contents

Part 1: Writing for Technical Communication

Part 2: Resources for Technical Communicators

Part 3: Roles of Technical Communicators

Introduction from the editor

Stephen Crabbe

Sixteen years have passed since the Institute of Scientific and Technical Communicators published its highly successful edited collection *Professional Communication and Information Design*. During this time, the need and demand for high quality technical and professional information has shown no signs of abating. I therefore felt that now was the right time for the Institute of Scientific and Technical Communicators to publish a new edited collection looking at current practices and trends in technical and professional communication. The result is this book, which brings together some of the leading academics and practitioners in the field of technical and professional communication from the UK and abroad. The book has been divided into three parts, with the first part 'Writing for Technical Communication' containing six chapters.

In Chapter 1, Lorcan Ryan defines the process of global authoring, explaining the key concepts associated with the process, examining those quality characteristics of relevance to global documentation, and describing the main techniques, resources and tools that make up a comprehensive global authoring strategy.

In Chapter 2, Mike Unwalla and Ciaran Dodd make the business case for ASD-STE100 Simplified Technical English (STE), showing how this controlled language is applicable in our current technology-driven world. In particular, they draw attention to how STE can give content clarity, consistency, structure and adaptability, and help make it easy to read and translate.

Readers need to evaluate writers and whether what they are reading is likely to be true. In Chapter 3, Kirstie Edwards argues in favour of self-representation in writing to help readers evaluate both the writer and the credibility of the writing.

In the next chapter (Chapter 4), Ellis Pratt suggests that technical writers should, from time to time, question whether the technical communication best practices of the past 25 years are still relevant today. He describes how some organizations are changing the way they write User Assistance in light of people's changing relationship with technology.

In Chapter 5, Andy Healey looks at the simplification of software and what it means for technical writers. He proposes that rather than technical documentation becoming obsolete, truly usable software assimilates it to be an intrinsic part of the product, as integral as any other element of the design.

In the final chapter in the first part of this book (Chapter 6), Klaus Schubert and Franziska Heidrich identify and discuss the categorization of barriers to communication, and the linguistic and communicational measures intended to enable text producers and recipients, respectively, to overcome, reduce or avoid those barriers, thus making content more accessible.

The second part of this book, titled 'Resources for Technical Communicators', contains five chapters. In the first chapter in this part (Chapter 7), Jody Byrne discusses key factors affecting the successful production of instructional videos on a large scale within an organization. These factors include practical tasks such as choosing the right software and equipment, as well as organizational issues such as staff engagement and adapting processes and workflows.

In Chapter 8, Marie Girard and Patricia Minacori argue that digital transformation brings a new level of complexity for technical communication. They suggest that systems thinking can address this challenge, in particular for auditing, governing and planning content, and adding flexibility and agility to the necessary structure of content systems.

In Chapter 9, Neal Goldsmith outlines the problems and obstacles that can be encountered when embarking upon an API documentation project and how to overcome them. He also provides guidelines on how to write good API documentation.

In Chapter 10, Keith Schengili-Roberts examines the history, development and growth of the DITA XML structured writing format. He provides information on who is using the standard and how its usage spreads from firm to firm. He then compares it to other structured authoring standards and reveals its likely future development.

In the final chapter of this part of the book (Chapter 11), Andrew McFarland Campbell shares his experiences with automating documentation to reduce the burden on technical writers when producing reference documents.

The third and final part of the book, titled 'Roles of Technical Communicators', contains five chapters. In the first chapter in this part (Chapter 12), Yvonne Cleary discusses trends affecting technical communication in Ireland. Her interviews with technical communicators, managers and company owners suggest that practitioners in Ireland work in agile environments, in virtual teams, and use new media tools to write and collaborate.

In Chapter 13, David Bird discusses the impact of information technology upon society, its collaborative effects, benefits and disadvantages, and his predictions as to its future.

In Chapter 14, Annette Wierstra and Joe Sellman argue that documentation reviews are not only about validating content but are also a tool to manage risks, and that using risk assessment to plan a review process that adequately weighs risks and project needs focuses the reviewer's attention where it is most needed.

Since 2009, University of Rennes 2 has been using the DITA standard to train students on structured writing principles and topic-based authoring. In Chapter 15, Nolwenn Kerzreho reveals how the DITA training has evolved and how a practical and project-based approach has proved to be the best option for the students at University of Rennes 2.

In the final chapter of the book (Chapter 16), Jason Lawrence and Chelsea Green explore the feasibility of Artificial Intelligence technical writers in the workplace. They review AI advancements and survey advanced AI technologies, including writing technologies. They conclude that AI writers are currently feasible and that the profession must develop best practices which include AI on writing teams.

I hope that you enjoy reading these chapters as much as I enjoyed putting them together. It is clear from this edited collection that the field of technical and professional communication is as vibrant and alive today as it was when *Professional Communication and Information Design* was published by the ISTC in 2001.

June 2017
Portsmouth

Part 1

Writing for Technical Communication

1

Global authoring: writing for a global audience

Lorcan Ryan

Abstract

Despite the paradigm of the English language functioning as a global *lingua franca* (Seidlhofer, 2005), the majority of the world's population is unable to speak, read or understand the English language (Lommel, 2003; Robert, 2005; Schäler, 2007; Internet World Stats, 2016). In the realm of digital content, the languid assumption that global information consumers *probably* speak and understand English has resulted in a situation where professional publishers of digital documentation have focused their efforts on optimizing content for native speakers of English, at the exclusion of the requirements of the other groups that make up a global audience.

Keywords
global authoring; global documentation; usability; readability; translatability

Introduction

Concentrating on composing content specifically for the needs of native English speakers is to ignore the fact that they only account for approximately 5% of the world's population (375 million people). This overlooks the other 1,125 million or so non-native speakers of English (15% of the world's population) who tend to use English merely as their chosen foreign 'contact language' rather than their preferred language of communication (Crystal, 2003; Seidlhofer, 2005), as well as the 5,900 million people (80%) who do not speak English at all.

In recent years, just as multinational corporations have attempted to manufacture culturally neutral 'global' products to sell in diverse international markets, digital content developers are striving to publish documentation in English that transcends lingua-cultural boundaries such that it is suitable for consumption by both native and non-native speakers, as well as being amenable to translation. We can refer to this type of documentation as 'global documentation' and the process of composing it as 'global authoring'.

This chapter will define global authoring and explain the key concepts associated with the process, including what we mean by the term 'global authoring', 'global audience', 'global author' and 'global documentation'. We will also cover the main quality characteristics of global documentation, and describe the main techniques, resources and tools that make up a comprehensive global authoring strategy.

Key concepts

There are several concepts that require clarification when discussing the process of writing for a global audience. To begin with, the term 'global' refers to anything that is independent of geographical, physical, and cultural contingencies (Wiegerling, 2004), and we can define 'global authoring' as the process of developing documentation that has been intentionally optimized for the requirements of a global audience. For the purposes of this discussion, it is necessary to narrow the scope of this definition somewhat. The source language that we will focus on, for example, will be English; and the term 'authoring' will refer to the creation of digital documentation rather than software applications, video games, or any other type of digital or non-digital content. In addition, we will place an emphasis on the processes involved in the composition of professionally produced digital documentation such as user assistance material, rather than text-based content produced by individuals for social reasons (such as personal blogs). Before discussing how to optimize this type of content for a global audience, it is necessary to explain what a global audience is comprised of.

In this chapter, we will use the term 'global audience' to refer to people dispersed throughout the world who speak and read in multiple different languages (Arnold, 1998; Brys and Vanderbauwhede, 2006), as opposed to a 'local audience' confined to a specific geographic area and who may only speak a single language. In terms of categorising the groups that make up a global audience, several approaches may be taken. The marketing department of a large multination corporation, for example, might segment their global customer base according to demographic characteristics such as social class, marital status, family size and spending power. However, in the context of global authoring, it may be more appropriate to classify the different groups that make up a global audience according to their proficiency in using the source language in which the documentation has been authored. In this way, we can adopt a similar approach to that taken by John Kohl (2008) among others (Dray and Siegel, 2006; Avval, 2011), who sub-divide a global audience into three main groups based on their usage of the English language:

- Native speakers (NS): Use English as their first language
- Non-native speakers (NNS): Speak English as a secondary language
- Non-speakers (NONS): Do not speak English at all

In theory, documentation that is genuinely *global* would transcend linguistic, cultural and geographic barriers and be universally understandable and acceptable. However, with the exception of a small percentage of documentation such as musical notation, it is practically impossible to author a single version of text-based content that is understandable and acceptable throughout the world due to the diverse languages and cultural norms that exist in different locales.

Given the difficulty in composing text-based content that is truly global in nature, a more viable strategy is to isolate and classify the groups that make up a global audience, identify which documentation quality characteristics are most relevant to these groups, and deploy the optimum mix of global authoring techniques, resources and tools to develop documentation that meets the requirements of each group. For the purposes of this chapter, we will adopt the classification approach mentioned earlier and categorize a global audience in terms of the proficiency of users with the English language. However, having already defined global authoring as the process of developing documentation that has been intentionally optimized for the requirements of a global audience… this raises an interesting dilemma. While we might envision optimizing English documentation for native and non-native speakers of the English language, how is it possible to optimize the content for the third group in our global audience: the non-speakers of English?

In truth, no matter how well-written English documentation is, if the target readers don't speak English, then we cannot consider that text as having been optimized for them without some level of translation or localization. This is where the notion of translatability comes into play.

Translatability refers to how easy or difficult it is to translate text from a source language into a given target language, and translatable documentation is content that is amenable to translation. Therefore, if we consider the third target group in our global audience to be the human translators who render the source language documentation into other languages for non-speakers of English rather than the non-speakers themselves, we can now classify the groups that make up our global audience more succinctly as follows: native speakers of the source language (NSs), non-native speakers (NNSs), and human translators (HTs). And following from this assumption, we can define 'global documentation' as text-based content that has been intentionally optimized to meet the requirements of native speakers and non-native speakers of the source language in which the documentation has been authored, as well as the human translators who render the documentation into different languages for non-speakers of the source language.

However, writing clear, concise content that is suitable for human translation, as well as consumption by native and non-native speakers, is a challenging process (Clark, 2009). We can collectively refer to professionals such as technical writers and instructional designers who are responsible for developing global documentation as 'global authors'. Professional authors are an essential part of an enterprise who are primarily responsible for selecting the appropriate mix of content (Albers, 2005), making it as engaging as possible (MacDonald, 2001), arranging and presenting it effectively for the relevant audience (Albers, 2005) and communicating it to end users (Bottita, 2003; Rockley, 2003). Albers (2003; 2005), Bottitta (2003) and Brys and Vanderbauwhede (2006) state that senior technical writers may have multiple responsibilities in addition to writing text, including constructing information models, performing audience analysis, developing style guides, incorporating usability and testing quality. Knowledge management (Applen, 2002) and optimization of page layout (Rascu, 2006) are other essential skills for professional authors, with Barr and Rosenbaum (2003) adding technical knowledge, organization and common sense to the list of authoring competences. And because global authoring refers to the process of developing documentation in such a manner that it is intentionally optimized for the requirements of a global audience, global authors should be familiar with the global documentation quality characteristics (GDQCs) most relevant to each of the three groups that make up a global audience (NSs, NNSs and HTs), as well as being well-versed in the global authoring techniques, resources and tools necessary to address these characteristics.

Despite researchers such as Barr and Rosenbaum (2003) suggesting that documentation quality should be a top priority for enterprises, studies have shown that up to 50% of all professional documents written by native speakers contain linguistic errors (Reuther, 1998). Therefore, there is clearly a need to improve the quality of professional documentation for end users. Quality, however, is a subjective element that is difficult to define in any context, let alone in relation to global documentation (Dilts, 2001; O'Brien, 2006; Gladkoff, 2010), and academics and practitioners often disagree on what constitutes quality in a document (Abbott and Eubanks, 2005). Secondary research reveals over 30 different quality attributes associated with digital documentation; from acceptability, clarity and completeness to consistency, motivation and navigation. For this purposes of this chapter, we will focus on GDQCs particularly relevant to the native speakers, non-native speakers and human translators that make up a global audience: usability, readability and translatability.

Human usability (referred to, in this chapter, simply as 'usability') is the first GDQC examined because it is recognized as one of the most important quality attributes of systems and documents, with consumers in 11 European countries ranking it second only to price as the most important product attribute (Schoeffel, 2003). Established researchers such as Jakob Nielsen describe usability as one of the most important quality attributes of user assistance (UA) documentation, regardless of whether the users are native or non-native speakers of the source language (Nielsen, 2000; Barr and Rosenbaum, 2003). In addition, it is widely accepted that usability encompasses several other quality attributes such as effectiveness, efficiency, and satisfaction (International Organization for Standardization (ISO), 1998; Frøkjær et al., 2000; Nielsen, 2003b; Sauro and Kindlund, 2005; Rubin and Chisnell, 2008; Doherty et al,. 2009).

Human readability (referred to, in this chapter, simply as 'readability') is the second GDQC analyzed because readable English documentation is not only important for native English speakers, but also for the non-native speakers (almost twice this number) who use English documentation, and the translators who render it into different languages (Hargis, 2000). Esselink (2000) opines that readability is the most important documentation quality attribute for non-native speakers, while Fry (2006) states that readability influences other quality attributes such as comprehension, retention, reading speed, reading persistence and reading enjoyment. As well as being a significant quality attribute for native and non-native speakers of a particular source language, several researchers believe that readability is also related to translatability (Means and Godden, 1996; Reuther, 1998; Barthe et al., 1999; Hargis, 2000; Bernth and Gdaniec, 2001).

Human translatability (abbreviated as HTran in this chapter) is the third GDQC that we will scrutinize because a key component of global documentation is its amenability to translation to make it understandable for non-speakers of the original source language in which it was authored. Researchers such as John Kohl (1999; 2008a) highlight its importance as a GDQC. We will place the focus on HTran rather than machine translatability (MTran), as it is even more under-researched than MTran, with no substantial body of associated research, and is therefore worthy of attention. In addition, MTran only refers to how suitable a text is for processing by machine-translation (MT) systems, and some of the rules that aim to make documentation more suitable for the processing requirements of a specific MT system may contradict the guidelines that aim to improve the quality of documentation for human readers. (For example, although long sentences are usually detrimental to the quality of MT output, human readers might not always prefer very short sentences over longer, more descriptive ones.) Therefore, using MTran as a GDQC may be a risky strategy as authoring documentation to meet the processing requirements of a specific MT system could contradict the requirements of a human reader or translator. Albrecht and Hwa (2008) also believe that the threat of a particular MT system becoming obsolete or discontinued is a valid concern if considering MTran as a GDQC.

Having explained the main concepts of global authoring, a global audience, global documentation and global authors, the next three sections in this chapter will explore the three selected GDQCs of usability, readability and translatability.

GDQC #1: usability

The first GDQC we will examine is usability. The term 'usability' was first used in the computer industry in the early 1980s, when global software companies such as IBM and Apple made concentrated efforts to ensure their products were accessible to all users, and not just specialists with technical expertise (Schoeffel, 2003). Usability is now recognized as an essential quality attribute of products, systems and documents (Schoeffel, 2003), with several specialized organizations offering usability consultancy services and an estimated 10,000 usability professionals employed worldwide (Nielsen, 2005b). Usability has also emerged as an academic discipline, with universities offering dedicated courses on usability and human-computer interaction. In addition, professional associations such as the Usability Professionals' Association (UPA) host usability conferences, publish usability papers, and promote industry standards and guidelines. Recognized authorities in the field of usability include:

- Jakob Nielsen (co-founder of the Nielsen Norman Group)
- Don Norman (co-founder of the Nielsen Norman Group)
- Bruce Tognazzin (co-founder of the Nielsen Norman Group)
- Jared Spool (founder of User Interface Engineering)
- Nick Finck (creator one of the first design and usability oriented online magazines, *Digital Web Magazine*)
- Janice Redish (founder of the Document Design Centre at the American Institutes for Research in 1979)

The International Organization for Standardization (ISO) (1998) defines usability as "the extent to which a product can be used by specified users to achieve specified goals with effectiveness, efficiency and satisfaction in a specified context of use." This definition has been developed by a panel of experts within a technical committee and, therefore, is a reasonably dependable source (International Organization for Standardization, 2014). Although the wording of the definition actually refers to product usability, researchers agree that effectiveness, efficiency and satisfaction are also crucial elements of documentation usability (Frøkjær et al., 2000; Nielsen, 2003b; Sauro and Kindlund, 2005; Nielsen, 2006, Rubin and Chisnell, 2008; Bevan, 2009; Doherty et al., 2009).

Effectiveness is the accuracy and completeness with which users achieve specified goals (International Organization for Standardization, 1998; Frøkjær et al., 2000; Rubin and Chisnell, 2008; Bevan, 2009), and its primary indicators are task completion rates and error rates (Frøkjær et al., 2000; Sauro and Kindlund, 2005). *Efficiency* refers to the resources expended by users in relation to the accuracy and completeness with which they achieve goals (International Organization for Standardization, 1998; Frøkjær et al., 2000) and its primary indicator is task completion time (Frøkjær et al., 2000; Sauro and Kindlund, 2005; Rubin and Chisnell, 2008; Bevan, 2009). *Satisfaction* is a gauge of users' attitudes, perceptions, feelings, and opinions toward a product (or document), and their comfort in using it (Lindgaard, 1994; International Organization for Standardization, 1998; Frøkjær et al., 2000; Rubin and

Chisnell, 2008). The primary indicator of satisfaction is the mean satisfaction rating of product users (Sauro and Kindlund, 2005). Nielsen (2001) and Rubin and Chisnell (2008) agree with using task completion rates, error rates, task completion times and user satisfaction as the basic measures of effectiveness, efficiency and satisfaction.

Researchers advocate several methods of capturing data related to documentation usability including:

- Usability tests (both informal usability tests (Jeffries et al., 1991; Nielsen and Levy, 1994; Kohavi et al., 2007; Nielsen, 2007) and controlled experiments)
- Heuristic evaluation (Keirnan et al., 2002; Nielsen, 2005c; The Usability Body of Knowledge, 2016)
- Cognitive walk-through (Nielsen and Levy, 1994)
- Fast iteration (Nielsen and Levy, 1994)
- Think-aloud protocol (Jeffries et al., 1991; Nielsen, 2005a; Doherty et al., 2009; The Usability Body of Knowledge, 2016)
- System Usability Scale (SUS) (Sauro and Kindlund, 2005; Bangor et al., 2009)

Controlled experiments provide one of the more robust means of evaluating usability because they employ a recognized empirical research methodology under meticulous conditions (Jeffries et al., 1991; Kohavi et al., 2007). Usability testing under controlled conditions reveals more usability issues by utilizing user observation (Karat et al., 1992), and generates more reliable data than more limited techniques such as heuristic evaluation (The Usability Body of Knowledge, 2016) and the System Usability Scale (SUS) (Bangor et al., 2009). Despite sometimes criticized for being expensive, cumbersome, and only providing a gloss after market release (Lindgaard, 1994), observing small numbers of live users attempt to complete scenario-based tasks in a controlled environment is a method advocated by respected usability researchers such as Jakob Nielsen (Nielsen, 2001; Nielsen, 2007). This evaluation method is also used in real-world projects by multinational corporations such as Oracle (Sauro and Kindlund, 2005).

The two main types of personnel involved in usability tests are the test facilitator and the participants. Researchers suggest including varying numbers of participants in usability tests. Participants should be representative of typical users of the product, system or document, and be familiar with what is being testing without having experience using the specific item itself (Gomoll and Nicol, 1990; Szafron and Schaeffer, 1994; Dilts, 2001; Keirnan et al., 2002; Faulkner, 2003; Nielsen, 2003a; Turner et al., 2006a). They should be randomly chosen from the relevant population to ensure that the usability test has a true experimental design (Rubin and Chisnell, 2008).

Usability evaluation methods such as controlled experiments and usability tests provide an opportunity to generate data about the usability of global documentation, which in turn may be used as a barometer to ascertain how effectively global authors are optimizing documentation for a global audience. The next section in this chapter will examine a second GDQC: readability.

GDQC #2: readability

The second GDQC we will analyze is readability. Readability is a measure of how easy or difficult a text is to read in terms of the complexity and ambiguity of its vocabulary (Means and Godden, 1996). Recognized authorities in the field of readability include:

- Rudolf Flesch (who created the Flesch Reading Ease test and was co-creator of the Flesch-Kincaid readability test in 1975)
- George Klare (a researcher best known for his research on assessing the grade level of texts for different readers)
- Edgar Dale (co-creator of the Dale and Chall Original Formula in 1948)
- Jeanne Chall (co-creator, along with Edgar Dale, of the Dale and Chall Original Formula in 1948)
- Robert Gunning (creator of the Fog Index readability formula in 1952)

While early studies did not necessarily equate readability with reader understanding (by Klare, 1963; Hargis, 2000; DuBay, 2004), more recent research often considers readability as synonymous with comprehensibility (Crystal, 1992; Means and Godden, 1996; Hargis, 2000). Fry (2006) states that readability influences other quality attributes such as comprehension, retention, reading speed, reading persistence, and reading enjoyment. As well as being a significant quality attribute for native and non-native speakers of a source language, readability may also be related to translatability (Means and Godden, 1996; Reuther, 1998; Barthe et al., 1999; Hargis, 2000; Bernth and Gdaniec, 2001). Although Reuther (1998) believes that readability rules are a subset of translatability rules, Bernth and Gdaniec (2001) found that improving the translatability of a text actually reduced its readability. Readability is often confused with legibility, although the latter is concerned with typeface and document layout rather than comprehension and reading speed (Hargis, 2000; DuBay, 2004). Other elements of readability include reading speed, reader interest and reader perseverance (Harkins and Plung, 1982; DuBay, 2004).

To gain a consistent understanding of readability, Dale and Chall provide a useful definition:

> The sum total (including all the interactions) of all those elements within a given piece of printed material that affect the success a group of readers have with it. The success is the extent to which they understand it, read it at an optimum speed, and find it interesting.

> (Dale and Chall, 1949, p.5)

Although originally conceived over 60 years ago, this definition is one of the founding characterizations of readability developed by two of the recognized authorities in the field of readability. The definition also encompasses three dimensions of readability frequently cited in subsequent definitions: comprehension, reading speed, and reader interest. Although the Dale and Chall (1949) definition was constructed prior to the proliferation of digital content and specifically refers to printed material, I extend their definition to encompass both printed and electronic text in this study.

In terms of evaluating readability, there are several different approaches which we can categorize into three main approaches: reader estimation, text analysis and reader testing. Reader estimation simply involves asking professional authors to judge how readable a text is (Klare, 1974), while text analysis involves counting a number of pre-defined language elements (such as polysyllabic words or excessively long sentences) in a text, and using this data to infer how readable the text is. Readability formulas give a measure of how difficult a piece of text is to read based on a score derived from data such as sentence length and word complexity (Klare, 1974; Redish and Selzer, 1985; Redish, 2000; DuBay, 2004; O'Brien, 2006). Although originally developed to assess whether school textbooks were appropriate for children at a particular grade level (Redish, 2000), readability formulas have been used more recently by enterprises such as Acrolinx to measure the quality of technical documentation. There are now more than 30 types of core readability formulas (Klare, 2000), with several hundred variations derived from these core formulas (Redish and Selzer, 1985; Hargis, 2000).

Reader testing, on the other hand, involves observing reader behaviour, collecting metrics such as reading speed, reader attitudes, reader retention rates and reader comprehension levels, and using these metrics to infer the readability of the text being examined (Klare, 1974; Harkins and Plung, 1982; Davison and Kantor, 1982; Redish and Selzer, 1985; Chall, 1988; Bernth, 1997; Hargis, 2000; Cadwell, 2008; Coleman et al., 2010). Although more expensive to implement than techniques such as readability formulas, reader testing is ranked as one of the most desirable readability evaluation options, as it involves actual readers and, unlike readability estimation and text analysis, helps identify places where a document is inaccurate, incomprehensible or poorly organized (Redish and Selzer, 1985; Schriver, 1991; Redish, 2000; Schriver, 2000).

Readability evaluation methods such as reader estimation, text analysis and reader testing provide an opportunity to generate data about the readability of global documentation, which in turn may be used as an indicator of how effectively global authors are optimizing documentation for a global audience. The next section in this chapter will inspect another significant GDQC: HTran.

GDQC #3: human translatability

The third GDQC we will scrutinize in this chapter is HTran. HTran refers to how easy or difficult it is for a human translator to translate text from a source language into a given target language (Campbell and Hale, 1999; Wilss, 2001; Cadwell, 2008), based on its grammatical quality (Kumhyr et al., 1994; Bernth, 1997; Akis and Sisson, 2010). Recognized authorities in the field of translatability include:

- Teruko Mitamura (faculty member of the Language Technologies Institute in Carnegie Mellon University, involved in controlled language and MT projects including KANT: Knowledge-based Machine Translation)
- John Kohl (charter member of the Association for Machine Translation in the Americas, SAS Institute linguistic engineer specialising in global English, author of *The Global English Style Guide*)
- Sharon O'Brien (faculty member of the School of Applied Language and Intercultural Studies in Dublin City University, author of several

publications on the areas of controlled language, machine translation, post-editing and localization)

- Arendse Bernth (a researcher best known for her publications on natural language, documentation quality and MTranslatability)
- Jeffrey Allen (a researcher known for his early contributions on the concept of author memory)

Clear and unambiguous language makes it easier for translators to render a text into different languages (Akis and Sisson, 2010). Translatability is also affected by the linguistic (Catford, 1965; Ping, 1999), intralingual (Ebel, 1969; Nida, 1984; Newmark, 1988; Wilss, 2001; Allen, 2002; Collins, 2003; Aiwei, 2005; Aidewuni, 2008) and socio-cultural (Bassnett-McGuire, 1980; Aiwe, 2004) challenges of attempting to transfer the semantic meaning conveyed in a source text to that of a completely different target language. In addition, the competence of the translator has an impact on translatability (Campbell and Hale, 1999; Avval, 2011). Given the challenges involved in accurately translating text from one language into another, Kitamura (1999) and Aiwei (2005) believe that this makes it more appropriate to refer to the untranslatability of a text. Gdaniec (1994), Bernth and Gdaniec (2001) and Underwood and Jongejan (2001) refer to parameters that negatively impact on the translatability of a text as translatability indicators, while O'Brien (2005; 2006) gives them the more specific label of negative translatability indicators (NTIs).

Although Means and Godden (1996), Hargis (2000), Reuther (2003) and Cadwell (2008) advocate a relationship between readability and translatability, Bernth and Gdaniec (2001) dispute this claim. Hargis (2000) also believes these two quality attributes are distinct entities, but states that testing the translatability of a text requires an assessment of its readability. Kumhyr et al. (1994) assert that documentation produced under proven translatability guidelines is easier and less time-consuming to translate. Haller and Schutz (2001) also claim that the way that technical documentation is written has a significant impact on its readability and translatability, but they had not conducted any experiments to substantiate this assertion. Iser (1996), on the other hand, maintains that translatability aims primarily to improve comprehension.

There currently is no standard technique for evaluating the HTran of a text (Sichel, 2009), most likely because the majority of measures of translatability are necessarily subjective (Gdaniec, 1994). Despite this lack of a universally recognized evaluation approach, we can categorize the existing approaches to measuring HTran as follows:

■ Source text analysis (STA)

■ Human translatability estimation (HTE)

■ Translatability inference (TIN)

STA involves evaluating the translatability of a source language text by counting the elements that could cause difficulties to a human translator. Examples of these types of problematic segments, referred to as negative translatability indicators (NTIs) by O'Brien (2006), include polysyllabic words, complex noun phrases and lengthy sentences. A quantitative measure for translatability can be derived by analyzing data such as the number of NTIs in a formula called a translatability index (TI) (Campbell and Hale, 1999; Bernth and McCord, 2000; Underwood and Jongejan, 2001; O'Brien, 2006). The TI, then, represents the complexity of a given text, its suitability for translation processing, and the time and cost necessary to translate it by generating a numerical score for its translatability (Gdaniec, 1994; Kumhyr et al., 1994; Bernth and Gdaniec, 2001; Underwood and Jongejan, 2001; O'Brien, 2005). Unfortunately, all existing TIs such as the Logos TI (Gdaniec, 1994), the Translation Confidence Index (TCI) (Bernth and McCord, 2000), the PaTrans TI (Underwood and Jongejan, 2001), and the Confidence Index (O'Brien, 2006) only generate MTran scores that indicate how amenable a text is to processing by a particular MT system, rather than how easy or difficult it is for a human translator to render into a different language. In fact, one of the only systematic attempts to identify HTran issues in a source text would appear to be John Kohl's Syntactic Cues Strategy (SCS) (Kohl, 1999). SCS involves following a 10-step approach to edit problematic linguistic features of a text to improve its HTran (Kohl, 1999; O'Brien, 2006).

Human translatability estimation (HTE) involves asking professional translators to judge how easy or difficult it would be to translate a text into another given language. Fiederer and O'Brien (2009) caution, however, that although language professionals may yield valuable insight into the translatability of a text, the use of human judgement can lead to issues such as subjectivity, time, and cost. HTE is similar to the readability estimation technique described in the previous section, albeit that it measures HTran rather than readability.

The third approach to translatability evaluation is what we can refer to as translatability inference (TIN), where the translatability of a source text is inferred from the quality of a translated version of that text. Unlike TSA and HTE, TIN involves translating a source text using human translators, evaluating the quality of the translated text with native speakers of the target language (Birch et al., 2010) and inferring, from that quality rating, how translatable the original source text was. The technique can be expanded to assess what impact the application of guidelines has on the HTran of documentation. Say, for example, that an enterprise produces two versions of an online help topic in English: the original version and a version that was rewritten to conform to the rules of a style guide or controlled language (CL). If that enterprise wants to find out which version of the English text is easier to translate they could apply a set of content development guidelines to a set of documentation to create an edited version, translate both versions into a given target language using human translators, evaluate the quality of both translated versions, and infer, from the quality ratings, which source language version was more amenable to human translation.

Having briefly inspected usability, readability and translatability; we will now examine the various components of a global authoring strategy which are implemented to ensure the production of high-quality global documentation.

Global authoring strategy: techniques, resources and tools

Key to the development of high-quality global documentation is the effective implementation of global authoring strategy encompassing a variety of global authoring techniques, resources and tools. Global authoring techniques refer to those methods employed by global authors to create digital documentation optimized for native and non-native speakers of the source language in which the content has been composed, as well as the human translators who render it into different languages for non-speakers of the source language. Global authors may implement several different techniques to develop global documentation including high-level global authoring principles, controlled authoring, writing for translation, minimalism and topic-based authoring.

A vast variety of high-level global authoring principles have been published by researchers and practitioners, including recommendations on how to conduct audience research, create authoring teams, implement authoring processes and use authoring tools. Audience research involves creating profiles of target readers including how technical or linguistically proficient they are, what motivates or interests them, and what kind of information they require (Duffy et al., 1992; Horton, 1994; MacDonald, 2001; Tsuji and Yoshikazu, 2001; Loorbach et al., 2006). Other recommendations refer to building teams of multi-skilled global authors (Horton, 1994; Dumas and Redish, 1999), and designing a global authoring process with clear objectives, user-centred document design, early prototyping and continuous revision (Horton, 1994; Lindgaard, 1994; Dumas and Redish, 1999; Beu et al., 2001; Bailey et al., 2006; Nielsen, 2007; Rubin and Chisnell, 2008). Still other high-level global authoring principles advise selecting authoring software carefully (Horton, 1994), and using design templates and style sheets to help achieve consistent formatting in documentation (Rockley, 2003). Controlled authoring, on the other hand, is the process of applying a more specific set of predefined style, grammar, punctuation and terminology rules during document design to standardize its structure, reduce ambiguity, and increase consistency (Murphy et al., 1998; Ó Broin, 2009). The strategy aims to produce source text that is easier for non-native speakers to read and understand, although it may also result in content that is more translatable (Murphy et al., 1998; Ó Broin, 2009). Ultimately, controlled authoring guidelines ensure that global documentation is useful, accurate, relevant, consistent and complete for target readers (Shneiderman, 1992; December, 1994; Horton, 1994; DuBay, 2004; Bailey et al., 2006).

Writing for translation is a strategy that focuses on authoring text that is easy to translate from the very birth of a document (Iverson and Kuehn, 1998; Murphy et al., 1998; Adams et al., 1999; Esselink, 2000; Cobbold and Pontes, 2007; Sichel, 2009). While controlled authoring involves standardizing information structure and facilitating content reuse, writing for translation emphasizes producing culturally neutral documentation with an absence of stereotypes, metaphors, slang and humour (Adams et al., 1999; Barnum and Lee, 2006). Indeed, DuBay (2004) suggests that the one of the 'golden rules' of documentation writing, regardless of medium, is to use culture- and gender-neutral language. This requires global authors to be knowledgeable about different cultures to avoid causing offence to international users (Arnold, 1998). Another tactic to increase the translatability of documentation is to embed syntactic cues into source text for human translators (Kohl, 1999).

Two additional global authoring techniques are minimalism and topic-based authoring. Minimalism is a strategy that aims to reduce complexity for readers and overall word count for translators, by authoring content in an extremely economical manner (Carroll, 1998; Hargis, 2000; Sichel, 2009), parallelism (Allen, 1999; O'Brien, 2003; Fawcett, 2009; Harris, 2009). Topic-based authoring techniques such as chunking and single-sourcing aim to improve the comprehension of documentation via consistency of information, and facilitate translation via leverage and reuse of previously translated chunks (Scattergood, 2002; Rockley, 2003; Ó Broin, 2005; O'Brien, 2006; Pietrangeli, 2009).

Global authors may also utilize resources such as dictionaries, thesauruses, lexical databases, and target reader profiles to develop high quality global documentation. Two particularly useful global authoring resources are global style guides and controlled languages (CLs). Style guides, whether global or organization-specific (Reuther and Schmidt-Wigger, 2000; Haller and Schutz, 2001; Bright, 2005; Rascu, 2006), are rulebooks containing standards for writing content (Nübel, 2004). Style guides consist of rules for consistent spelling, grammar and formatting (Bright, 2005), and are particularly important for multi-authored works (Bergeron and Balin, 1997). Global style guides, such as *The Global English Style Guide* (Kohl, 2008), aim to improve readability and comprehension for non-native speakers and translators, as well as improving the quality of the content for native-speakers of the source language. Style guidelines are also published in CL rule sets. A CL is a subset of a natural language (O'Brien, 2006) that restricts the grammar, style and vocabulary of that language in order to reduce its ambiguity and complexity, and improve its consistency, readability, translatability and retrievability (Means and Godden, 1996; Reuther, 1998; Bernth and Gdaniec, 2001; Rascu, 2006; Muegge, 2009).

CLs constrain grammar and style to keep elliptical constructions, multiple coordinated sentences, conjoined prepositional phrases, relative clauses and ambiguous language to a minimum (Mitamura, 1999, Torrejon and Rico, 2003). Some controlled languages also incorporate a controlled vocabulary with a limited set of approved words with restricted meanings, a thesaurus of unapproved terms and suggested alternatives and guidelines for adding new technical words to the approved vocabulary (Bergeron and Balin, 1997; Mitamura, 1999; International Federation of Library Associations, 2005). Generic controlled languages, such as Basic English, ASD Simplified Technical English, Controlled Language Optimized for Uniform Translation (CLOUT) and Controlled Language Authoring Technology (CLAT), are sets of CL rules developed for generic content rather than the specific content developed by an individual organization (Nübel, 2004; O'Brien, 2006; Muegge, 2009; Basic-English Institute, 2010). Some enterprises have also developed proprietary CLs such as Caterpillar Technical English (CTE) and General Motors Controlled Automotive Service Language (CASL), to improve the quality of the global documentation that they publish

(Kamprath and Adolphson, 1998; Torrejon and Rico, 2003; Ó Broin, 2009).

In addition to global authoring techniques and resources, the successful implementation of global authoring tools also facilitates the creation of global documentation. In this chapter, we use 'global authoring tool' as a generic term to refer to those content-creation technologies capable of checking for text-based linguistic issues that may impact on the quality of documentation for a global audience. Although several types of software applications such as spell checkers, word processors and desktop publishing tools may incorporate basic style and grammar checkers, controlled language (CL) checkers offer a more robust solution to developing high-quality global documentation as they control both the vocabulary and terminology of a text as well as its style and grammar. CL checkers verify that source language text conforms to the rules of a specific CL using functions such as rule checking, term checking, batch checking, interactive checking and suggestions for word edits and sentence rewrites (Bernth, 1997; Murphy et al., 1998; Akis et. al., 2003). The main objective of CL checkers is to improve the consistency and translatability of documentation (Akis et. al., 2003; O'Brien, 2006), with Bernth (1997) stating that they also improve precision, generality, customisability and user-friendliness. Commercially available CL checkers such as Acrocheck™ and SDL AuthorAssistant™ may be customized to check text for a set of pre-defined linguistic issues such as sentences that exceed a certain word count, for example. Enterprises such as Sun Microsystems, Boeing, Bosch, General Motors, IBM, Motorola, Philips, SAS Institute, Siemens and Symantec have also developed their own proprietary internal CL checkers to improve the consistency, translatability and comprehensibility of the global documentation that they publish (Akis et al., 2003; Ó Broin, 2005; Kohl, 2008b; Roturier, 2009).

A comprehensive global authoring strategy, therefore, includes determining the optimum mix of global authoring techniques, resources and tools. To create high-quality global documentation, global authors should be proficient with techniques such as implementing high-level global authoring principles, controlled authoring, writing for translation, minimalism and topic-based authoring; have access to resources such as dictionaries, thesauruses, lexical databases, target reader profiles, global style guides and CLs; and be trained on how to use global authoring tools such as CL checkers. The next section will summarize the main points of writing for a global audience that we have covered in this chapter.

Projections for the future

In terms of the future, publishers of English source documentation are beginning to consider the requirements of a global audience, which includes non-native speakers of the English language and human translators, as well as native English speakers. Large multinational corporations are creating roles for global authors, investing in global authoring tools and building databases of locale-specific information to assist with the process of composing culturally neutral content.

The commercial benefits of developing high-quality global documentation include increased customer satisfaction, and the reduced costs associated with translation and localization (Reuther, 1998; Reuther, 1998; Torrejon and Rico, 2003; Rico and Torrejon, 2005). Although the investment level associated with implementing a global authoring strategy may be initially expensive, new technologies such as CL checkers may provide efficiencies to offset this cost.

As a new trend in the world of technical communication, global authoring looks set to continue to grow in importance. Key areas of growth may lie in continuing to develop global technologies to optimize content for a global audience, evolving evaluation techniques and metrics for GDQCs such as translatability, and developing a universally accepted framework of guidelines with which to script high-quality global documentation.

References

Abbott, C. and Eubanks, P. (2005) How academics and practitioners evaluate technical texts: A focus group study. *Journal of Business and Technical Communication*, 19 (2), 171–218.

Adams, A., Austin, G., and Taylor, M. (1999) Developing a resource for multinational writing. *Technical Communication*, 46 (2), 249–254.

Adewuni, S. (2008) Linguists and culture experts at a crossroad: Limitations in formulating an experimental translation theory. *Translation Journal*, 12 (2).

Aiwei, S. (2005) Translatability and poetic translation. Available at http://www.translatum.gr/journal/5/translatability-and-poetic-translation.htm [accessed 15 Dec 2016].

Akis, J.W. and Sisson, W.R. (2010) Improving translatability: A case study at Sun Microsystems. Available at http://www.lisa.org/globalizationinsider/2002/12/improving_trans.html [accessed 22 Feb 2012].

Akis, J.W., Brucker, S., Chapman, V., Ethington, L., Kuhns, B., and Schemenaur, P.J. (2003) Authoring translation-ready documents: Is software the answer? *Proceedings of SIGDOC '03*, New York: ACM, 12–15 Oct, 39–44.

Albers, M. (2005) The future of technical communication: Introduction to this special issue. *Technical Communication*, 52 (3), 267–272.

Albrecht, J. and Hwa, R. (2008) Regression for machine translation evaluation at the sentence level. *Machine Translation*, 2 (1–2), 1–27.

Allen, J. (1999) Adapting the concept of 'translation memory' to 'authoring memory' for a controlled language writing environment, *Proceedings of Translating and the Computer 20*, London: ASLIB, 10-11 Nov.

Allen, J. (2002) The Bible as a resource for translation software. *MultiLingual Computing & Technology*, 13 (7), 8–13.

Applen, J. (2002) Technical communication, knowledge management and XML, *Technical Communication*, 49 (3), 301.

Arnold, M. (1998) Building a truly World Wide Web: A review of the essentials of international communication. *Technical Communication*, 45(2), 197–206.

Avval, S. (2011) How to avoid communication breakdowns in translation or interpretation? *Translation Journal*, 16 (2).

Bailey, R.W., Barnum, C., Bosley, J., Chaparro, B., Dumas, J., Ivory, M.Y., John, B., Miller-Jacob, H. and Koyani, S.J. (2006) *Research-based web design and usability guidelines*. Washington D.C: U.S. Government Printing Office.

Bangor, A., Kortum, P., and Miller, J. (2009) Determining what individual SUS scores mean: Adding an adjective rating scale. *Journal of Usability Studies*, 4 (3), 114–123.

Barnum, C. and Li, H. (2006) Chinese and American technical communication: A cross-cultural comparison of differences. *Technical Communication*, 53 (2), 143–165.

Barr, J. and Rosenbaum, S. (2003) Documentation and training productivity benchmarks. *Technical Communication*, 50 (4), 471–484.

Barthe, K., Juaneda, C., Leseigneur, D., Loquet, J., Morin, C., Escande, J. and Vayrette, A. (1999) GIFAS Rationalized French: A Controlled Language for aerospace documentation in French. *Technical Communication*, 46 (2), 220–229.

Basic-English Institute (2010) Rules of grammar. Available at http://ogden.basic-english.org/rules.html [accessed 03 July 2016].

Bassnett-McGuire, S. (1980) *Translation Studies*, London: Methuen & Co. Ltd.

Bergeron, B. and Bailin, M. (1997) The contribution of hypermedia link authoring. *Technical Communication*, 44, 121–128.

Bernth, A. (1997) EasyEnglish: A tool for improving document quality. *Proceedings of ANLC '97*, 159–165.

Bernth, A. and Gdaniec, C. (2001) MTranslatability. *Machine Translation*, 16, 175–218.

Bernth, A. and McCord, M. (2000) The effect of source analysis on translation confidence: Envisioning machine translation in the information future. *Proceeedings of the 4th conference of the Association for Machine Translation in the Americas*, AMTA 2000, Berlin: Springer Verlag.

Beu, A., Hassenzahl, M. and Burmester, M. (2001) Engineering joy. *IEEE Software*, 18 (1), 70.

Bevan, N. (2009) International standards for usability should be more widely used. *Journal of Usability Studies*, 4 (3), 106–113.

Birch, A., Osborne, M., and Blunsom, P. (2010) Metrics for MT evaluation: Evaluating reordering. *Machine Translation*, 24, 15–26.

Bottitta, J., Idoura, A., and Pappas, L. (2003) Managing the effects of organizational changes, *Technical Communication*, 50 (3).

Braster, B. (2009) Controlled language in technical writing. *MultiLingual Computing & Technology*, 20(1).

Bright, M. (2005) Creating, implementing, and maintaining corporate style guides in an age of technology. *Technical Communication*, 52 (1).

Brys, C. and Vanderbauwhede, W. (2006) Communication challenges in the W3C's web content accessibility guidelines. *Technical Communication*, 53 (1).

Cadwell, P. (2008) Readability and controlled language: Does the study of readability have merit in the field of controlled language, and is readability increased by applying controlled-language rules to texts? Unpublished MA thesis, Dublin City University.

Campbell, S. and Hale, S. (1999) What makes a text difficult to translate? *Proceedings of the 23rd Annual ALAA Congress*, 19 April.

Carroll, J.M. (1998) *Minimalism beyond the Nurnberg Funnel*. Cambridge, Massachusetts: MIT Press.

Catford, J.C. (1965) *A linguistic theory of translation: An essay on applied linguistics*. London: Oxford University Press.

Chall, J. (1988) The beginning years. In: Zakaluk B. (Ed.) and Samuels, S.J. (Ed.), *Readability: Its past, present and future*. Newark: International Reading Association.

Cobbold, G.L. and Pontes, R. (2007) Five steps from local to global. *MultiLingual Computing & Technology*, October/November 2007.

Coleman, C., Lindstrom, J., Nelson, J., Lindstrom, W., and Gregg, K.N. (2010) Passageless comprehension on the Nelson-Denny Reading Test: Well above chance for university students. *Journal of Learning Disabilities*, 43 (3).

Collins, A. (2003) Complying with European language requirements. *MultiLingual Computing & Technology*, 14 (3).

Crystal, D. (1992) *An encyclopaedic dictionary of language and languages*. Massachusetts: Blackwell.

Crystal, D. (2003) *English as global language*. Cambridge University Press.

Dale, E. and Chall, J.S. (1949) The concept of readability. *Elementary English*, 26 (23).

Davison, A. and Kantor, R. (1982) On the failure of readability formulas to define readable texts: A case study from adaptations. *Reading Research Quarterly*, 17 (2), 187–209.

Dilts, D. (2001) Successfully crossing the language translation divide. SIGDOC '01 *Proceedings of the 19th Annual International Conference on Computer Documentation*, Sante Fe, New Mexico, USA, October 21–24, 2001, 73–77.

Doherty, G., Van Der Sluis, I., Karamanis, N. and Luz, S. (2009) Designing personalised, multilingual speech and language based applications. Available at http://citeseerx.ist.psu.edu/viewdoc/download?doi=10.1.1.228.1380&rep=rep1& type=pdf [accessed 30 Aug 2016].

Dray, S.M. and Siegel, D.A. (2006) Melding paradigms: Meeting the needs of international customers through localisation and user-centred design. In K.J. Dunne (Ed.) *Perspectives on localisation*. Amsterdam/Philadelphia: John Benjamins.

DuBay, W.H. (2004) The principles of readability. Available at http://www.impact-information.com/impactinfo/readability02.pdf [accessed 11 Mar 2016].

Duffy, T.M., Mehlenbacher, B. and Palmer, J.E. (1992) *Online help: Design and evaluation*. Norwood, New Jersey: Ablex Publishing Corporation.

Dumas, J.S. and Redish, J.C. (1999) *A practical guide to usability testing*. London: Intellect Books.

Ebel, J.G. (1969) Translation and cultural nationalism in the reign of Elizabeth. *Journal of the History of Ideas*, 30, 593–602.

Esselink, B. (2000) *A practical guide to localization*. Amsterdam: John Benjamin Publishing Co.

Faulkner, L. (2003) Beyond the five-user assumption: Benefits of increased sample sizes in usability testing. *Behaviour Research Methods, Instruments and Computers*, 35(3), 379–383.

Fawcett, H. (2009) Effective rhetoric, effective writing: Parallelism in technical communication. Available at https://helenfawcett.com/technical-writing/effective-rhetoric-effective-writing-parallelism-in-technical-communication/ [accessed 17 Jun 2016].

Fiederer, R. and O'Brien, S. (2009) Quality and machine translation: A realistic objective? *Journal of Specialised Translation*, 11.

Frøkjær, E., Hertzum, M. and Hornbæk, K. (2000) Measuring usability: Are effectiveness, efficiency, and satisfaction really correlated? *Proceedings of the ACM Conference on Human Factors in Computing Systems (CHI) 2000*, Washington D.C: ACM Press, 345–352.

Fry, E. (2006) *Readability*. Newark: International Reading Association.

Gdaniec, C. (1994) The Logos translatability index. *Proceedings of the First Conference of the Association for Machine Translation in the Americas*, AMTA, 97–105.

Gladkoff, S. (2010) Language quality assurance: The business, the science, the practice and the tool. *TCWorld Magazine*, Nov 2010. Available at http://www.tcworld.info/e-magazine/translation-and-localization/article/language-quality-assurance-the-business-the-science-the-practice-and-the-tool/ [accessed 30 May 2016].

Gomoll, K. and Nicol, A. (1990) Discussion of guidelines for user observation. User Observation: Guidelines for Apple Developers.

Haller, J. and Schutz, J. (2001) CLAT: Controlled Language Authoring Technology. SIGDOC'01, October 21–24, 2001, Santa Fe, New Mexico, USA.

Hargis, G. (2000) Readability and computer documentation. *ACM Journal of Computer Documentation*, 24 (3).

Harkins, C. and Plung, D.L. (1982) *A guide for writing better technical papers*. New York: IEEE Press.

Harris, R. (2009) A handbook of rhetorical devices. Available at http://www.virtualsalt.com/rhetoric.htm [accessed on 03 Jul 2016].

Horton, W. (1994) *Designing and writing online documentation: Hypermedia for self-supporting products*, 2nd Ed. New York: John Wiley and Sons.

International Federation of Library Associations (2005) Guidelines for multilingual thesauri. Available at http://archive.ifla.org/VII/s29/pubs/Profrep115.pdf [accessed 11 Feb 2016].

International Organization for Standardization (1998) *9241-11: Guidance for Usability*, 1st Ed. Geneva, Switzerland: International Organization for Standardization.

International Organization for Standardization (2014) How does ISO develop standards? Available at http://www.iso.org/iso/home/standards_development.htm [accessed 26 Oct 2016].

Iser, W. (1996) On translatability. Available at http://www.pum.umontreal.ca/revues/surfaces/vol4/iser.html [accessed 7 Jul 2016].

Iverson, S. and Kuehn, H. (1998) Assessing translation readiness: A maturity level. Available at www.gbv.de/dms/goettingen/247341444.pdf [accessed 19 Mar 2016].

Jeffries, R., Miller, J.R., Wharton, C., and Uyeda, K.M. (1991) User interface evaluation in the real world: A comparison of four techniques. *Proceedings of the ACM Conference on Human Factors in Computing Systems (CHI) 1991*, 119-124, New Orleans, Louisiana: ACM Press.

Kamprath, C., Adolphson, E., Mitamura, T. and Nyberg, E. (1998) Controlled language for multilingual document production: Experience with Caterpillar Technical English. *Proceedings of the Second International Workshop on Controlled Language Applications (CLAW '98)*, Pittsburgh.

Karat, C., Campbell, R. and Fiegel, T. (1992) Comparison of empirical testing and walkthrough methods in user interface evaluation. *Proceedings of the ACM Conference on Human Factors in Computing Systems (CHI) 1992*, 397-404, Monterey, California: ACM Press.

Keirnan, T., Anschuetz, L. and Rosenbaum, S. (2002) Combining usability research with documentation development for improved support. ACM Special Interest Group (SIG) on the Design of Communication (DOC) SIGDOC'02, October 20-23, Toronto, Canada.

Kitamura, K. (2009) Cultural untranslatability. *Translation Journal*. Available at http://translationjournal.net/journal/49translatability.htm [accessed 01 Jun 2016].

Klare, G. (1963) *The measurement of readability*. Ames, Iowa: Iowa State University Press.

Klare, G. (1974) Assessing readability. *Reading Research Quarterly*, 10 (1), 62-102.

Klare, G. (2000) The measurement of readability: Useful information for communicators. *ACM Journal of Computer Documentation*, 24 (3).

Kohavi, R., Henne, R., and Sommerfield, D. (2007) Practical guide to controlled experiments on the Web: Listen to your customers not to the HiPPO. ACM SIG Data Mining Knowledge Discovery'07, ACM.

Kohl, J. (1999) Improving translatability and readability with syntactic cues. *Technical Communication*, 46 (2), 149–166.

Kohl, J. (2008a) *The Global English style guide: Writing clear, translatable documentation for a global market.* New York: SAS Institute Inc.

Kohl, J. (2008b) Language quality-assurance software: Optimising your documentation for a global audience. *Intercom*, 55 (5).

Kumhyr, D., Merrill, C. and Spalink, K. (1994) Internationalization and translatability. *Proceedings of the First Conference of the Association for Machine Translation in the Americas*, Association for Machine Translation in the Americas, Washington, D.C., USA.

Lindgaard, G. (1994) *Usability testing and system evaluation: A guide for designing useful computer systems.* London: Chapman and Hall.

Loorbach, N., Steehouder, M., and Taal, E. (2006) The effects of motivational elements in user instructions. *Journal of Business and Technical Communication*, 20(2), 177–199.

MacDonald, M.P. (2001) Can a manual entertain? *Intercom: The Magazine of the Society for Technical Communication*, June 2001, 14–17.

Means, L. and Godden, K. (1996) The Controlled Automotive Service Language (CASL) project. *Proceedings of the First International Workshop on Controlled Language* (CLAW).

Mitamura, T. (1999) Controlled language for multilingual machine translation. *Proceedings of Machine Translation Summit VII*, Singapore, September 13–17.

Muegge, U. (2009) Controlled language – does my company need it? Available at http://www.tcworld.info/e-magazine/content-strategies/article/controlled-language-does-my-company-need-it [accessed 24 Apr 2016].

Murphy, D., Mason, J. and Sklair, S. (1998) Improving translation at the source. Translating and the Computer 20, *Proceedings from AsLib conference*, 12–13 November 1998.

Newmark, P. (1988) *A textbook of translation.* Hertfordshire: Prentice Hall.

Nida, E.A. (1984) *On translation.* Beijing: China Translation & Publishing Corporation.

Nielsen, J. (2000) The mud-throwing theory of usability. Available at http://www.useit.com/alertbox/20000402.html [accessed 24 Jun 2016].

Nielsen, J. (2001) Usability metrics. Available at http://www.useit.com/alertbox/20010121.html [accessed 24 Jun 2016].

Nielsen, J. (2003a) Recruiting test participants for usability studies. Available at http://www.useit.com/alertbox/20030120.html [accessed 24 Jun 2016].

Nielsen, J. (2003b) Usability 101: Introduction to usability. Available at http://www.useit.com/alertbox/20030825.html [accessed 24 Jun 2016].

Nielsen, J. (2005a) Authentic behaviour in user testing. Available at http://www.useit.com/alertbox/20050214.html [accessed 24 Jun 2016].

Nielsen, J. (2005b) Usability for the masses. *Journal of Usability Studies* 1, 2–3.

Nielsen, J. (2005c) How to conduct a heuristic evaluation. Available at http://www.useit.com/papers/heuristic/heuristic_evaluation.html [accessed 24 Jun 2016].

Nielsen, J. (2006) Quantitative studies: How many users to test? Available at http://www.useit.com/alertbox/quantitative_testing.html [accessed 24 Jun 2016].

Nielsen, J. (2007) Fast, cheap, and good: Yes, you can have it all. Available at http://www.useit.com/alertbox/fast-methods.html [accessed 24 Jun 2016].

Nielsen, J. and Levy, J. (1994) Measuring usability: Preference vs. performance. *Communications of the ACM*, 37 (4).

Nübel, R. (2004) Evaluation and adaptation of a specialised language checking tool for non-specialised machine translation and non-expert MT users for multi-lingual telecooperation. Available at www.mt-archive.info/LREC-2004-Nuebel.pdf [accessed 26 Sep 2016].

Ó Broin, U. (2005) Taking care of global business: Translatability and localization of e-business applications. *MultiLingual Computing & Technology*, July/August 2005.

Ó Broin, U. (2009) Controlled authoring to improve localization. *MultiLingual Writing for Translation, Getting Started: Guide*, October/November 2009.

O'Brien, S. (2003) Controlling Controlled English: An analysis of several controlled language rule sets. *Proceedings of the EAMT/CLAW 2003*, Dublin, Ireland.

O'Brien, S. (2005) Methodologies for measuring the correlations between post-editing effort and machine translatability. *Machine Translation*, 19 (1), 37–58.

O'Brien, S. (2006) Machine translatability and post-editing effort: An empirical study using Translog and choice network analysis. Unpublished PhD thesis, Dublin City University.

Pietrangeli, L. (2009) Internationalising your content: Authoring with localization in mind. TC World Magazine. Available at http://www.tcworld.info/e-magazine/translation-and-localization/article/internationalizing-your-content-authoring-with-localization-in-mind/

Ping, K. (1999) Translatability vs. untranslatability, *BABLE*, 45 (4), 289–300 [accessed 07 Jul 2016].

Rascu, E. (2006) A controlled language approach to text optimisation in technical documentation. *Proceedings of KONVENS*, 107–114.

Redish, J.C. (2000) Readability formulas have even more limitations than Klare discusses. *ACM Journal of Computer Documentation*, 24(3), August 2000.

Redish, J.C. and Selzer J. (1985) The place of readability formulas in technical communication. *Technical Communication*, 32(4), 46–52.

Reuther, U. (1998) Controlling language in an industrial application. *Proceedings of the Second International Workshop on Controlled Language Applications*, May 1998, Pennsylvania: Carnegie Mellon University.

Reuther, U. (2003) Two in one: Can it work? Readability and translatability by means of controlled language. *Proceedings of EAMT/CLAW 2003: Controlled Language Translation*, Dublin, Ireland.

Reuther, U. and Schmidt-Wigger, A. (2000) Designing a multi-purpose CL application. *Proceedings of the Third International Workshop on Controlled Language Applications*, 72–82, CLAW 2000, Seattle.

Rico, C. and Torrejon, E. (2005) Translation technology in Spain: The Observatorio de Tecnologias de la traduccion, *Localisation Focus*, 4 (2).

Rockley, A. (2003) Managing enterprise content: A unified content strategy. Available at http://www.rockley.com/articles/The%20Rockley%20Group%20-%20ECM%20UCS%20Whitepaper%20-%20revised.pdf [accessed 24 Aug 2016].

Roturier, J. (2009) Deploying novel MT technology to raise the bar for quality: A review of key advantages and challenges. *MT Summit XII: proceedings of the twelfth Machine Translation Summit*, August 26–30, Ottawa, Ontario, Canada.

Rubin, J. and Chisnell, D. (2008) *Handbook of usability testing*, 2nd Ed., New York: John Wiley and Sons.

Sauro, J. and Kindlund, E. (2005) Making sense of usability metrics: Usability and Six Sigma. *Proceedings of the Usability Professions Association (UPA) Conference 2005*, Montreal, Quebec: UPA.

Scattergood, D. (2002) Documentation localisation costs. *Localisation Focus*, 1(2).

Schoeffel, R. (2003) The concept of product usability. *ISO Bulletin 2003*, 34(3), 6–7.

Schriver, K.A. (1991) Plain language for expert or lay audiences: Designing text using user edit. *Technical Report Number 46*, Pittsburgh, PA: Carnegie Mellon University, Communications Design Center.

Schriver, K.A. (2000) Readability formulas in the new millennium: What's the use? *ACM Journal of Computer Documentation*, 24(13).

Shneiderman, B. (1992) *Designing the user interface: Strategies for effective human-computer interaction*, 2nd Ed., New Jersey: Addison-Wesley.

Sichel, B. (2009) Planning and writing for translation. *MultiLingual Writing for Translation, Getting Started: Guide*, October/November 2009.

Szafron, D. and Schaeffer, J. (1994) An experiment to measure the usability of parallel programming systems, *Concurrency Practice and Experience*.

The Usability Body of Knowledge (2016) Methods: Heuristic evaluation. Available at http://www.usabilitybok.org/heuristic-evaluation [accessed 15 Feb 2016]

Torrejon, E. and Rico, C. (2003) Controlled languages in the localization industry. *Localisation Focus* 2(4).

Tsuji, S. and Yoshikazu, Y. (2001) A framework to provide integrated online documentation. *Proceedings of the 19th Annual International Conference on Computer Documentation*, 185–192, New York: ACM.

Turner, C., Lewis, J. and Nielsen, J. (2006) Determining usability test sample size. *International Encyclopedia of Ergonomics and Human Factors*, 2nd Ed., Volume 3.

Underwood, N. and Jongejan, B. (2001) Translatability checker: A tool to help decide whether to use MT. *Proceedings of MT Summit VIII*, 363–368, Santiago de Compostela, Galicia, Spain, 18–22 Sep.

Wiegerling, K. (2004) Localisation versus globalisation: Claim and reality of mobile and context-aware applications of the Internet. *International Centre for Information Ethics* (ICIE), 2, 1–7.

Wilss, W. (2001) *The science of translation – Problems and methods.* Shanghai: Foreign Education Publishing House.

2

The case for ASD-STE100 Simplified Technical English

Mike Unwalla and Ciaran Dodd

Abstract   This book shows the challenges that face technical communicators today. We are a long way from conventional paper manuals, linear documents and static illustrations. Now we have video tutorials, graphic-only instructions, mobile technology and adaptive content. Do text-based tools like ASD-STE100 Simplified Technical English® (ASD-STE100) still have a role? Our answer is emphatically yes.

In Part 1 of this chapter, we explain what ASD-STE100 is, how it works, the benefits of using it and how it is relevant today. In Part 2, we discuss the challenges of implementing ASD-STE100.

Keywords   ASD-STE100 Simplified Technical English; clarity; controlled language; safety-critical documentation; translatability

Introduction

The aim of technical communication is to communicate meaning clearly and concisely. Readers must be able to find the information that they want and understand it in the way that the technical communicator intended. Although many things have changed in technical communication in the last few years, this purpose remains the same.

Since its creation in 1986, the controlled language ASD-STE100 has been a valuable tool to achieve this purpose. Although ASD-STE100 has been available for 30 years, in our experience few people know about

ASD-STE100, and those that do have limited understanding and often misconceptions.

We think that ASD-STE100 deserves a chapter in this book for three reasons:

1 ASD-STE100 tells technical communicators how to write clear, consistent, easy-to-understand technical documentation for both native and non-native English readers.
2 Despite the changes of the last 30 years, ASD-STE100 is still relevant. It is also adaptable to industries outside aerospace.
3 When technical documentation is written in ASD-STE100, technical communicators, organizations and readers all benefit.

Part 1: The evolution of ASD-STE100

What is ASD-STE100?

In the early 1980s, the international aerospace community examined the readability of maintenance documents. The output of these studies was AECMA Simplified English, which was released in 1986. In 2005, the AeroSpace and Defence Industries Association of Europe (ASD) adopted AECMA Simplified English and renamed it *ASD-STE100 Simplified Technical English: International specification for the preparation of maintenance documentation in a controlled language* (ASD-STE100).

The current issue is ASD-STE100 issue 7 from January 2017 (ASD, 2017). In the specification, ASD-STE100 is also referred to as STE. To show that the specification is applicable to all industries, the title was changed. The specification is now for the preparation of 'technical documentation', not only 'maintenance documentation'. ASD-STE100 is maintained by the Simplified Technical English Maintenance Group (STEMG), a group of representatives from various organizations who make sure that ASD-STE100 remains up-to-date. To learn more about the history of ASD-STE100, or to get the ASD-STE100 specification, refer to the STEMG website (STEMG, 2016a).

ASD-STE100 is a controlled language. It restricts the complexity of the text by controlling:

■ the grammar of the language; and
■ the words that you can use.

Part 1 of the specification has 53 writing rules in nine sections. Part 2 of the specification contains the dictionary of approved words.

The nine sections of rules specify:

- What types of words you can use (three categories: approved words from the dictionary, technical names and technical verbs)
- How to structure sentences and paragraphs
- How to write procedures, descriptions and safety instructions.

"The basic principle of STE is to make texts easy to read and understand" (ASD, 2017, p. 1-4-1). Together, the writing rules and the dictionary help writers to remove ambiguity. The basic principles are that one word has one meaning, and that text must be consistent and specific (rather than abstract). This before-and-after example from the specification shows these principles (ASD, 2017, p. 1-7-2).

✗ Incorrect:

CAUTION: EXTREME CLEANLINESS OF OXYGEN TUBES IS IMPERATIVE.

✓ Correct:

WARNING: MAKE SURE THAT THE OXYGEN TUBES ARE FULLY CLEAN. OXYGEN AND OIL OR GREASE MAKE AN EXPLOSIVE MIXTURE. AN EXPLOSION CAN CAUSE DEATH OR INJURY TO PERSONNEL AND/OR DAMAGE TO EQUIPMENT.

In this example, the original text is incorrectly labelled a caution (damage to equipment). It is an abstract statement that does not clearly describe the danger, or what the reader must do to be safe. This example also shows that ASD-STE100 is not about shortening sentences. It is about making technical documentation as clear as possible.

Learning to write well in ASD-STE100 is like learning a new type of English rather than simplifying standard English. ASD-STE100 is not like baby talk or pidgin English. For example, you cannot write 'Power switch ON' in ASD-STE100, because you omit important words that help the reader to make sense of the text. This is different to some traditional technical writing. The old military specification MIL-M-81927B(AS) says that if space is limited, you can omit articles and verbs if the meaning will be clear to readers (Eveland, 1990). ASD-STE100 has rules that require you to include important words like articles (Rule 2.3) or that prohibit the omission of words (Rule 4.2). To get the benefits of ASD-STE100, you must learn both the correct vocabulary and the grammatical rules.

As this quote from the STEMG shows (STEMG, 2016a, p. FAQ), ASD-STE100 is not applicable to all types of document:

> AECMA developed STE to improve the procedures and descriptive text in aerospace maintenance documents… STE can improve maintenance documentation in other industries, but it is possible that some documents (for example, those which are analytical, too descriptive, or legalistic in nature) will not get this same benefit.

In the next sections, we explore what we mean by a controlled language, how controlled languages work and how ASD-STE100 compares to other forms of simpler English.

Controlled languages

A controlled language is a natural language such as French, Spanish, or Chinese that has limits on how grammar and words are used (ISO, 2015). Most controlled languages have two parts (Muegge, 2009):

- A set of grammar rules
- A set of approved terms.

Examples of controlled languages other than English are Simplified Technical Spanish (Remedios, 2003) and ScaniaSwedish (Almqvist and Hein, 1995). We will focus on controlled English.

Kuhn (2013) lists at least 100 versions of controlled English. Many versions are proprietary, for example, from Avaya, Caterpillar, and Sun Microsystems (Muegge, 2008), and it is difficult to get information about them.

Most controlled languages are used for technical texts, but EasyEnglish is a controlled language for religious texts (Betts, 2005). We will compare the rules of ASD-STE100 and EasyEnglish to show how controlled languages work.

ASD-STE100 rules

These examples from the ASD-STE100 dictionary (Figure 1) show how the writing rules and the dictionary of approved terms work together to control language.

The ASD-STE100 dictionary has approximately 880 approved terms. Usually, a term is permitted for only one part of speech. For example, the term *oil* is approved as a noun:

✖ Incorrect: You must oil the valve.

✓ Correct: You must put oil on the valve.

The rules for writing specify the structure of the text. For example, for an instruction write "a maximum of 20 words in each sentence" (ASD, 2017, p. 1-5-1).

ASD-STE100 specifies the permitted verb tenses (Rule 3.2): infinitive, imperative, simple present, simple past, and future. Also, the past participle is permitted as an adjective. The *ing* form is not permitted unless it is part of a technical name, or in a title to describe a task (Rule 3.5).

✖ Incorrect: While removing the oil filter...

✓ Correct: When you remove the oil filter...

Some people think that ASD-STE100 requires technical communicators to use only the words in the dictionary, but this is not correct. Each organization has its own special vocabulary, sometimes with many thousands of technical terms. The specification has categories of technical names (Rule 1.5) and technical verbs (Rule 1.12). If you can put a noun or a verb into a category, the word will qualify as a technical name or a technical verb and so become approved. Thus, if organizations have different technical terms for the same thing, like *safety goggles* or *safety eyewear*, ASD-STE100 can accommodate these differences. These rules are, in part, what makes ASD-STE100 adaptable to industries outside aerospace.

EasyEnglish rules

EasyEnglish (Betts, 2005) is an example of a non-technical controlled language, but it works in a similar way to ASD-STE100.

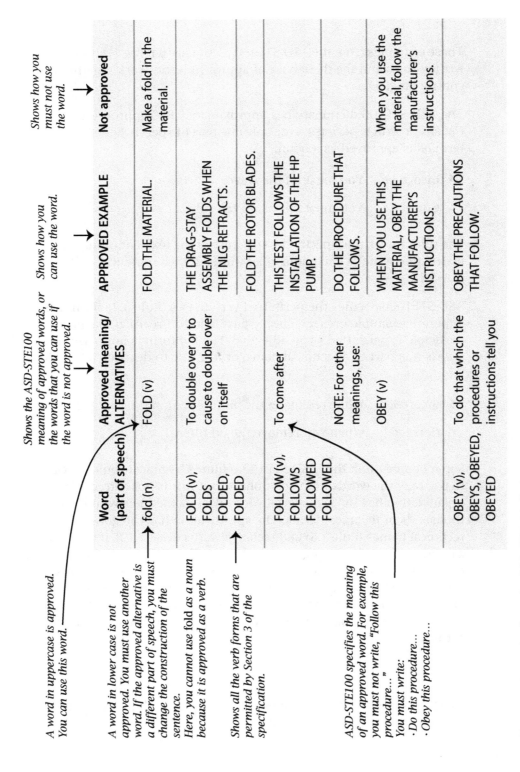

Figure 1: Examples from the ASD-STE100 dictionary

EasyEnglish uses three methods to make text clear:

- It restricts vocabulary
- It uses simple grammar
- It applies a logical structure to make text as clear as possible.

EasyEnglish has two levels of vocabulary. Level A has approximately 1200 words and is applicable to people who have a working knowledge of English as a second language. Level B has approximately 2800 words and is applicable to people who have an intermediate level of proficiency in English.

The rules of EasyEnglish remove some of the problems that the rules of ASD-STE100 also aim to remove. For example, the first languages of some readers of EasyEnglish have few abstract nouns, thus EasyEnglish avoids abstract nouns such as *peace* and *strength*. Some languages do not use the passive voice, and speakers of those languages struggle to understand the English passive voice. If possible, EasyEnglish does not use the passive voice. In ASD-STE100, the passive voice is prohibited in procedures and is used "only when it is really necessary" in descriptions (ASD, 2017, p. 1-3-5).

Like ASD-STE100, EasyEnglish has rules to keep sentences short and simple. For example, use a maximum of one subordinate clause in a sentence (the examples are ours):

✖	Incorrect:	Make sure that the carpet that is in the bedroom is clean.
✓	Correct:	The carpet is in the bedroom. Make sure that the carpet is clean.

Simple words and simple grammar are not sufficient to make text clear, thus EasyEnglish applies a logical structure to change standard English to EasyEnglish:

1 Identify all of the ideas that are in a sentence.
2 Put the ideas into a logical sequence.
3 Write a sequence of EasyEnglish sentences.

This is similar to ASD-STE100 Section 4, which says "...you must write short sentences and use [a] simple sentence structure" (ASD, 2017, p. 1-4-1).

Other forms of simpler English

The term *simplified* in ASD-STE100 Simplified Technical English can cause confusion because other types of English are simpler than standard English. STEMG (2016b) writes:

> It must be clear that ASD-STE100 does not intend to compete with other forms of Simplified English, Plain English, Simple English or any other Controlled Language and must not be confused with them...

These other types of simpler English are not necessarily designed for technical communications, but it is useful to compare them with a controlled English like ASD-STE100. Some other types of simpler English are:

- Basic English, Ogden (1940)
- Globish, Nerrière and Hon (2009)
- Plain English
- Global English, Kohl (2008).

We will discuss plain English and Kohl's Global English.

Plain English

Plain English, unlike ASD-STE100, is more a set of principles for writing clearly than a carefully documented set of rules like those in ASD-STE100. Unlike ASD-STE100's STEMG, no central authority defines and governs plain English. Instead, different organizations champion the use of plain English and sell plain English services and approvals. There is no standard against which to validate a text for conformity to plain English.

In the United States of America, the U.S. Securities and Exchange Commission (SEC, 1998) wrote:

> A plain English document uses words economically and at a level the audience can understand. Its sentence structure is tight. Its tone is welcoming and direct...A plain English document is easy to read and looks like it's meant to be read.

In the United Kingdom, the Plain English Campaign (PEC, 2016a) states, "Since 1979, we have been campaigning against gobbledygook, jargon and misleading public information...We believe that everyone should have access to clear and concise information." Their definition of plain

English is "a message, written with the reader in mind and with the right tone of voice, that is clear and concise" (PEC, 2016b).

We studied the SEC (1998) and Plain English Campaign (PEC, 2016b) guidance documents, and compared them with ASD-STE100. Table 1 compares the guidance.

Table 1: Plain English compared with ASD-STE100

Plain English SEC 1998	*Plain English Campaign*	*ASD-STE100*
Use short sentences	Use short sentences (average 15–20 words)	Procedural text 20-word maximum Descriptive text 25-word maximum
Replace jargon and legalese with short, common words	Use familiar language for your audience	Language is specified in the rules in Section 1: Words
Try personal pronouns	Use personal pronouns	Some personal pronouns are permitted
Write in the active voice as much as possible: A did B (active voice) B was done by A (passive voice)	Write in the active voice as much as possible	Use only the active voice in procedural writing Use the active voice as much as possible in descriptive writing
Use verbs not nouns made from verbs: Make *payment* (noun) by 6 June. *Pay* (verb) by 6 June	Use verbs not nouns made from verbs	Use verbs

Plain English does not give guidelines about how to write for people who read English as a second language. However, helping non-native English speaking readers was one of the primary reasons that ASD-STE100 was created. "The *one word one meaning* approach is much easier for non-native speakers to understand than plain English because all ambiguity is removed from the language" (Eveland, 1990).

For example, multi-word verbs cause problems for people who read English as a second language:

- *Carry out* means *do*
- *Put up with* means *tolerate*
- *Work out* means *calculate*.

Frequently, the meaning of a multi-word verb is different from the meanings of each word in the verb. ASD-STE100 Rule 9.3 tells writers not to combine words in this way. Thrush (2001) wrote, "Even very advanced learners of English have not mastered these idiomatic expressions."

In summary, we think that the primary differences between plain English and controlled English are as follows:

- In plain English, all words and grammatical structures are permitted unless a guideline says "do not use this". With a controlled language, a technical communicator must not use a word or a grammatical structure unless it is approved.
- Plain English is not a specification. Different practitioners can have different opinions about whether text is plain English.
- Although there is much agreement between plain English and ASD-STE100, the cohesion of the rules, the approved words and the underlying principles make the guidance in ASD-STE100 much clearer.

Global English

Another form of simpler English is Kohl's Global English (Kohl, 2008), which tries to make technical documentation as clear as possible, both for translation and for readers.

Do not confuse Kohl's Global English with other types of 'global English' or with 'international English'. Some basic principles are the same, but two important differences exist:

- Kohl gives hundreds of detailed technical guidelines.
- All the guidelines are based on research at SAS (www.sas.com).

Kohl (2008, p. 14) writes, "Global English could be regarded as a loosely controlled language," but:

> In the development of Global English, the emphasis has been on identifying grammatical structures and terms that should be *avoided*, rather than cataloguing structures and terms that are *allowed*. In other words, anything that is not specifically prohibited or cautioned against is allowed.

Global English is more restrictive than plain English, but it is less restrictive than ASD-STE100. As with plain English, Global English has rules in common with both ASD-STE100 and plain English. Global English guidelines restrict the use of the passive voice and specify a simple style of writing. But, the Global English guidelines are more detailed than the guidelines for plain English.

For example, one guideline is to use syntactic cues. A *syntactic cue* is a part of language that helps a reader to identify parts of speech and to analyze the structure of a sentence. Sometimes, syntactic cues are optional, but a sentence that does not have syntactic cues can be ambiguous. Kohl (2008, p. 119) gives a humorous example. The grammatically correct sentence, "Do not dip your bread or roll in your soup" has two interpretations:

✓ Correct: Do not dip your bread or your roll in your soup.

✓ Correct: Do not dip your bread in your soup, and do not roll in your soup.

If readers understand the second interpretation, they know that although it is grammatically correct, it is nonsense. But, with technical texts, if a technical communicator does not include optional syntactic cues, a reader's interpretation can be incorrect with possibly dangerous results.

As these examples show, to use Global English well, you must know English grammar well. This has an implication for organizations that want to adopt Global English. In our experience, many technical communicators have a sense of what is grammatically correct, but would find the level of detail in the Global English rules overwhelming.

To summarize, many differently named types of simpler English each have different levels of control. Although each type of English is the product of its purpose, they share many common principles, and all try to convey meaning clearly and concisely.

Thirty years of ASD-STE100 and changes in technical communication

In 2002, Ciaran began training technical communicators in ASD-STE100. Usually, her clients were in the aerospace industry and had to use ASD-STE100. The technical communicators worked in large teams. They produced technical documentation, usually in English and not for translation, that was published in paper format or in PDF files. Some technical communicators used specialist authoring software rather than Microsoft Word. In the wider world, mobile meant laptops. The Internet was not as sophisticated as it is today and mobile technology was in its infancy.

Today, Ciaran still trains clients from the aerospace industry but increasingly her clients are from other industries like the automotive, oil and gas, agricultural and semi-conductor industries. These clients choose to use ASD-STE100. The technical communicators work in very small teams, some with colleagues all over the world. Some technical communicators still write, but others curate content or check documentation that is outsourced to technical documentation organizations. The technical communicators write in many languages that are translated and published in many outputs for many types of device. The technical communicators who write in English possibly speak English as a second language. Many technical communicators use structured writing and content management systems. The concept of content strategy has emerged to further challenge the traditional world of technical communication.

Before we examine the challenges, we consider who else uses ASD-STE100. Braster (2008) gives these examples:

- The Credit Union Central of British Columbia (CUCBC) is the trade association and central banker for 48 independent credit unions in British Columbia. The *CUCBC Operations Manual* is an important product that is supplied to the credit unions of British Columbia. CUCBC used ASD-STE100 to standardize the manual, and to make sure that the quality conformed to legal and regulatory requirements.

- Elekta is a medical technology group that supplies systems and services to improve cancer care and the management of brain disorders. Elekta adopted ASD-STE100 to increase safety, to improve the quality of its technical documentation and to decrease translation costs. It customized ASD-STE100 to create its own controlled language.

- WatchGuard Technologies manufactures security products for computer networks. WatchGuard Technologies adopted ASD-STE100 to improve readability and reduce translation costs. It customized ASD-STE100 to create its own controlled language.

Kaiser (2016) states that 58% of requests for the ASD-STE100 specification come from industries other than aerospace and defence. Thus, if organizations are *choosing* to use ASD-STE100, it must still be relevant in today's world.

In the last 30 years, technical communication has changed in the following ways:

- Technology – there have been many changes both in how technical communicators produce and publish technical information and in how readers access the technical information: mobile technology, augmented reality, apps, video, and verbal assistants like Siri.

- Adaptive content – providing content for readers when, where, and how they want it.

- Collaboration and removing silos – the boundaries between the people who create technical content and those who use it have blurred, especially with the Internet and social media. For example, technical content may be used in marketing or for customer support and readers post help and queries online so that they become both readers and producers of technical content.

- Globalization – technical communicators are frequently remote from their readers and colleagues, physically, linguistically and culturally. For example, content is authored in English by non-native English speakers or in languages that are translated. Organizations localize content to different markets.

- Standardization and process – readers expect to have the same experience of an organization wherever they are in the world and when they do not, their satisfaction with the organization decreases. To meet this expectation, organizations standardize their content and the way that they produce it.

- Structure – for reuse, ease of maintenance and searchability.

- Content strategy – recognizing that technical content is more than a necessity but is an asset that should be controlled, managed, used and archived.

For examples of some of these changes, read the keynote presentation from the TCUK conference in 2013 (Rolls-Royce, 2013). At the conference, the speakers from Rolls-Royce plc explained how their publications evolved and their plans for the future.

The speakers described the problems of producing paper documentation, the move to using CDs and then the adoption of an electronic publication system in the 1990s. With this publication system coming to the end of its life and the current need to use data more effectively, the speakers described the planned global publications system. Based on S1000D (ASD, 2012), the new system will integrate data from various parts of the organization to support the maintenance operations. The presentation shows how complex global organizations like Rolls-Royce are adapting to make the most of their valuable data.

In another example of change, Avrahami (2016) describes how readers seek information from "a variety of digital touchpoints...across a mix of devices. Unfortunately for consumers, most organizations fail at delivering consistent experiences across all touchpoints." The article gives reasons why this happens. The relevant part of the article for us has suggestions for achieving consistency "across all touchpoints":

1 Create "componentized content for reuse across multiple deliverables. Reusing content helps ensure accuracy, consistency, improves usability, reduces cost and supports the delivery of personalized content."

2 "Adopt content creation tools that guide [all content creators] toward the creation of content that is consistent in structure and terminology."

3 "Classify and semantically tag all product content regardless of which department created it...to automatically deliver content to those that need it."

All Avrahami's suggestions require content to be controlled, structured and consistent, which is what writing in ASD-STE100 achieves. ASD-STE100 is still relevant, especially because in the last few issues the STEMG has revised the rules and terminology to make the specification less focused on aerospace.

The reasons for using ASD-STE100

Makes texts easier to read

ASD-STE100 makes texts easier to read by:

- Controlling the words that you can use
- Including the simplest words in the dictionary
- Specifying that one word has one meaning or one part of speech
- Emphasising the need to be consistent
- Simplifying the grammar that you can use.

Safety

ASD-STE100 originated in the aerospace industry, where English is the working language, but many readers of the maintenance documents are not native English speakers. The airlines wanted clearer instructions in the maintenance documents so that technicians could do the correct maintenance "to guarantee the aircraft availability". The Service information letter / Position paper states, "Complex technical instructions can be misunderstood and misunderstandings can lead to accidents. STE... [is a] valuable resource...to remove linguistic barriers and reduce Human Factors risks" (STEMG, 2016b).

The United States Federal Aviation Administration (FAA, 2014) wrote:

> ...good documents must have the right content, the right readability, and the right organization...Readability means that the procedure needs to use unambiguous grammar and terms, and have diagrams that are designed for the ultimate user. For example, Simplified Technical English is a proven way to reduce comprehension errors.

Doherty and Chiarello (2014) discussed whether using a controlled language, specifically ASD-STE100, in operating and maintenance manuals would improve safety in the marine industry. The authors concluded that procedures in a controlled language reduce human error and that ASD-STE100 is flexible enough for the marine industry.

Requirement

In aerospace and defence, ASD-STE100 is specified or recommended in these standards (STEMG, 2016a, p. About):

- ATA iSpec 2200 (was ATA 100)
- ATA 104 Training
- S1000D
- European Defence Standards Reference (EDSTAR).

Common language and consistency

Usually, technical communicators get input from subject matter experts (SMEs) and their documents are approved either by the same SMEs or by management. Each person has a preferred style of writing. Without a standard, these differences in style become apparent either in the document or in how people review and correct the document. If everyone complies with ASD-STE100, there is a common language for all people who produce, review and read the documents.

Structure and discipline at the sentence level

Structured writing has become an established feature of technical communication. Structured writing focuses on structuring documents so that the content is better categorized, structured, organized and labelled:

> The term structured writing was first used by Robert E. Horn, inventor of the Information Mapping Method. Structured writing is the process of creating and systematically organizing independent units of information so they are easily accessible, manageable and reusable.
>
> (Information Mapping, 2016)

> Structured writing is the act of creating content that obeys one or more predefined and explicitly recorded constraints that serve a defined purpose, in a format readable by machines, with the goal of making the content better.
>
> (Baker, 2015)

Much structured writing is done with XML and related technologies.

A key term used in structured writing is *granularity*. Granularity refers to the level of detail contained in each chunk of information. But, structured writing does not specify how to write these chunks of information at sentence and paragraph level. Because ASD-STE100 controls the language that we write, the process of constructing text in ASD-STE100 puts structure into text at paragraph and sentence level. The rules in Section 6 about descriptive writing show this, for example, "Paragraphs organize a text into logical units, and help to hold the reader's attention" (ASD, 2017, p. 1-6-7) and "The example has short paragraphs, which give the text a clear structure and make it easy for the reader to understand" (ASD, 2017, p. 1-6-5).

Translations – increase quality and decrease cost

Documents written in ASD-STE100 should not need translating if readers have a "basic knowledge of English". Without this knowledge of English, readers will need translated documents and ASD-STE100 helps the translation process (STEMG, 2016a, p. FAQ).

Translation is one of the easier ways to quantify the benefits of writing in ASD-STE100. According to Braster (2008), content can be decreased by up to 20% by deleting the 'nice to know' information. (Deleting 'nice to know' information echoes John Carroll's concept of minimalism in technical communication. See Hackos (2012) for more detail.)

Writing the source text in ASD-STE100 removes inconsistencies and standardizes text. Thus, if words or sentences were translated before and are stored in a translation memory, the organization will not need to pay for these words to be translated again. But, if a component has different names, each name will need to be translated, which increases the cost and possibly creates confusion.

Translation costs can be decreased by up to 40% for each language (Braster, 2008). The cost of translation depends on the number of words that are translated and the ease with which the source language translates to the target language. The discipline of writing in ASD-STE100 increases the quality of the source text and decreases the likelihood of errors in the target language.

Part 2: ASD-STE100 in practice

Implementing ASD-STE100

In Ciaran's experience, there are two approaches to implementing ASD-STE100.

1 Focus only on ASD-STE100 and train the technical communicators to use the specification. Often an organization adopts this approach because it is contractually required to use ASD-STE100 or wants to improve the quality of its content, especially for non-native English speaking readers or for translation.

2 Include ASD-STE100 as part of a larger change in authoring processes such as moving to structured writing. Sometimes, organizations will also buy an ASD-STE100 checker.

Can you successfully use ASD-STE100 without training? In Ciaran's experience, people who are asked to write in ASD-STE100 without training usually focus only on the words in the dictionary. This loses much of the effectiveness of ASD-STE100. Ciaran agrees with the STEMG (2016a, p. Training) who write, "STE training is the first essential step for a technical author to be able to apply STE correctly."

Training

For training in ASD-STE100, the STEMG recommends their authorized training provider or training from a member of the STEMG (STEMG, 2016a, p. Training). But, the STEMG recognizes that there are other training providers, and on the same web page, it gives some questions to help businesses to choose a training provider.

Some training providers deliver courses on ASD-STE100 only. Some academic and technical communication courses include modules on controlled languages or ASD-STE100. Some providers of ASD-STE100 checkers include training on ASD-STE100. The sources and types of training are many and varied, so businesses can choose training that suits their budgets, the learning styles of the delegates, the time available and the number and location of delegates that need training. For example, some businesses and delegates prefer the immediacy of face-to-face classroom training to allow delegates to question the trainer and get immediate feedback.

To use the rules of ASD-STE100 correctly, you must know English well and understand basic grammar. During training, it is useful to revise the parts of grammar necessary to apply the rules because some adults were not taught grammar and others learned grammar a long time ago. To write well in ASD-STE100, you must make conscious your unconscious knowledge of English.

Unless an organization *must* use ASD-STE100, to get approval for training is often difficult. Often, the need for ASD-STE100 is championed by one technical communicator, or the technical publications manager, who must persuade senior management and budget holders to fund the training. Some typical problems are:

- ASD-STE100 and its benefits are not always understood or valued by people other than technical communicators.
- ASD-STE100 is not usually a priority for budget holders.
- The logistics are difficult or expensive because organizations have one or two technical communicators, often in offices spread geographically, so it is difficult to justify the cost of classroom training.

But, if the training occurs, usually technical communicators can see the value of ASD-STE100 as a tool for writing. The training can highlight some of the challenges of implementing ASD-STE100, the subject that we discuss next.

The challenges of implementing ASD-STE100

Terminology management

ASD-STE100 permits an organization to include its own technical terms. Ideally, as a minimum, an organization should record:

- Approved nouns, approved adjectives, and approved verbs (technical names and technical verbs)
- Synonyms of the approved terms
- Examples of correct text and incorrect text.

The reality is that organizations, especially large global organizations, rarely have the time or the resources to define their terms. Ciaran knows of only two organizations that have formally reviewed and defined their terms, one of which was a smaller organization with a team dedicated to implementing ASD-STE100.

Even if you cannot review all of the terms used in your organization, it is beneficial to manage the terms used on a specific project. For example, Mike used this process on one project:

1 Collect an initial set of terms from these sources:
   - The primary contractor's list of preferred terms
   - The customer's style guide
   - The procedures that the primary contractor sent to the customer as part of the bid.
2 Discuss the initial list of terms with SMEs and the customer to identify the approved terms and the alternatives that are not approved.
3 Receive the first draft of a procedure from a technical communicator.
4 From the first draft and from the related source documents, identify candidate technical terms and the not approved equivalent terms. The Internet is a good source of information, especially for the names of organizations and regulations.
5 Discuss the technical terms with the technical communicator to find the best terms to use.
6 Send the technical terms to the customer for approval or correction.
7 Edit the procedure to conform to ASD-STE100 as much as possible.
8 Send the procedure to the technical communicator for the next review.

9 Put all the approved terms and the not approved terms into Microsoft Excel. The technical communicators had access to Excel through SharePoint.

10 Add the approved terms and the not approved terms into the ASD-STE100 checker.

It is important to involve the SMEs in this process. In this project, text contained the term *intrinsically safe*. Initially, Mike removed *intrinsically*, because synonyms of intrinsically are *basically*, *by nature*, *essentially*, *fundamentally*, *inherently*, and *innately*. But, the SMEs explained that *intrinsically safe* is a technical adjective.

Make sure that there is a procedure for change control. During the project, the customer changed some previously approved terms, which caused inconsistency in the titles of related documents and checklists.

Give the reviewers sufficient guidance to do their job. In the project, Mike did not give sufficient guidance to the reviewers, so some reviewers changed good English into bad English and approved terms to not approved terms.

Resistance from technical communicators

Some technical communicators resist ASD-STE100 for these reasons:

- ASD-STE100 restricts creativity by standardising text. For example, the recommendation in ASD-STE100 that American English spelling is preferred to British spelling often provokes a negative response. (But, ASD-STE100 permits British English.)
- ASD-STE100 is not a familiar style of writing. On one course, a delegate said that he preferred *obtain* to the permitted word *get*.
- ASD-STE100 is too simplistic or repetitive, especially if the technical communicator learned a literary style at school and is not trained to write in a controlled language (Eveland, 1990, p. 40).
- ASD-STE100 is too blunt. One person felt uncomfortable using the word *obey*, because it is more forceful than the unapproved word *follow*.

Even the most positive technical communicators will leave a training course with concerns about the practicalities of adopting ASD-STE100:

- It takes time to be proficient in ASD-STE100 but delegates are rarely given time to build their confidence after the course.

- Delegates are frequently concerned about the workload immediately after the course and about how they will check their work, especially without an ASD-STE100 checker.

- To write well in ASD-STE100, you must structure your content at paragraph and sentence level. This is difficult when technical communicators must produce documents quickly and often without all of the necessary data.

- It is much easier to write new content in ASD-STE100 than to change existing text into ASD-STE100. The reality is that technical communicators often copy and paste content from existing materials so they have to convert the text into ASD-STE100.

- Frequently, technical communicators get information from SMEs who do not know ASD-STE100. The technical communicators must then get the same SMEs to approve their technical content. The technical communicators are squeezed between SMEs who do not understand ASD-STE100 and customer-facing colleagues who want the material quickly.

Many of these challenges come from people's natural resistance to change, which can be managed. The same organization that formed a project team to implement ASD-STE100 conducted a series of roadshows to explain to their colleagues about their ASD-STE100 implementation. This organization is unusual. Another global organization followed up their training in ASD-STE100 with regular webinars for their technical communicators (on different continents) to compare progress on their implementations, and with an annual conference. Other organizations have trained the SMEs and management in ASD-STE100 so that they can better support the implementation. There are solutions to managing resistance, but they must be planned and resourced.

ASD-STE100 checkers

You can write ASD-STE100 correctly without software (STEMG, 2016a, p. Software), but software is useful. Although the ASD-STE100 grammar rules are simple, the specification contains many thousands of terms. It is not easy to make sure that the text conforms to the rules about terminology and grammar.

To use an ASD-STE100 checker effectively, you must customize it to include approved technical terms and optionally, the not approved technical terms. "To be effective, they [ASD-STE100 checkers] also need to contain your in-house technical names and technical verbs. If your technical names and technical verbs are not in the checker, you will get constant 'unknown word' messages" (STEMG, 2016a, p. Software).

These examples show why customization is necessary:

- In ASD-STE100 issue 7, the word *base* is unapproved as a noun and as a verb (but it is approved as a technical name). You have an approved technical name *base deck*.
- The term *root directory* has the synonyms *home directory* and *root*.
- An alternative for the adjective *5-part* is *five-part*.
- Some organizations use SI units of measure (litres, metres), and some organizations use non-SI units of measure (gallons, feet).

Although an ASD-STE100 checker can help you, it does not replace the skill of a technical communicator. You must not use the suggestions from the software without thinking.

Evaluation of ASD-STE100 checkers

Evaluate an ASD-STE100 checker fully before you buy it. Specify your requirements and then compare the software against the requirements. Refer to STEMG (2016a, p. Software) for a checklist of questions and to the *Software Evaluation Guide* (Jackson, Crouch, Baxter, 2014).

Ideally, at least one person who evaluates the ASD-STE100 checker must have a good knowledge of ASD-STE100. Without that knowledge, your team will struggle to know whether the ASD-STE100 checker finds all the errors and ignores all the correct text. Most ASD-STE100 checkers are likely to give incorrect warnings. Part of your evaluation is to decide whether the frequency of incorrect warnings is sufficiently low.

Problems of analysis

Some ASD-STE100 rules are easy to automate. For example, the word *abandon* is unapproved. An ASD-STE100 checker can easily find all instances of the term *abandon* and its inflections (*abandons*, *abandoned*, *abandoning*). Technical names can include an unapproved term. After you add the technical name to the software, the ASD-STE100 checker can ignore the unapproved term if it is part of a technical name.

Other rules are more difficult for an ASD-STE100 checker to analyze. For example, "Make sure that each paragraph has only one topic" (Rule 6.5).

Sometimes, an ASD-STE100 checker gives an incorrect analysis because the rules for the analysis of text are not sufficiently good. (We use the term *rule*, but possibly, some ASD-STE100 checkers use statistical methods to analyze text.)

Sometimes, an analysis is not possible because a text can be parsed in two ways. For example, in ASD-STE100 the passive voice is not permitted for a procedure. "The wires are disconnected" has two grammatical interpretations:

- The sentence is in the passive voice.
- *Disconnected* is an adjective that specifies the condition of the wires. (Compare with, "The wire is dirty.")

The term *operating system* is approved as a technical name (ASD-STE100 Rule 1.5.19), but the verb inflection *operating* is not approved. With "Operating systems that are slow can cause problems", the term can be analyzed in two ways:

- ✗ Verb + noun (not correct ASD-STE100): *Operating* systems that are slow can cause problems. (= If you operate systems that are slow, problems can occur.)

- ✓ Technical name (correct ASD-STE100): *Operating systems* that are slow can cause problems. (= If an operating system is slow, problems can occur.)

Sometimes, an analysis is difficult because a term is approved only when it has a particular meaning. For example, with ASD-STE100, the word *about* is approved only if it has the meaning *concerned with*. You must not use the word *about* to mean *approximately*.

Word-sense disambiguation is a difficult problem. People use their knowledge about the world, to disambiguate meaning. For example, think about the sentence, "She sat by the bank in the warm sunshine". Without context, you cannot know whether *bank* means a building or a grassy area at the side of a river. But, if you read the sentence in a story about summer, romance, daffodils, and true love, you know that *bank* almost certainly means the place near a river.

A simple ASD-STE100 checker can only find a word and ask, "Did you mean...?" But, Boeing developed the Boeing Meaning-Based Checker to give better analyzes (Remedios, 2003).

Some ASD-STE100 rules can be analyzed only in structured texts. For example, Rule 5.3 tells you to use the imperative form in an instruction.

Thus, in XML, if the sentence is marked as a step, there must be a verb in the imperative form. But, if text has no structure, an ASD-STE100 checker cannot 'know' whether a sentence is an instruction.

Some XML authoring tools integrate with ASD-STE100 checkers.

```
7 ▾  ··<body>↵
8    ····<p>The·examples·that·follow·are·from·the·'Not·approved'·column·in·the·
     ASD-STE100·specification:</p>↵
9 ▾  ····<ol>↵
10   ······<li>Actuate·the·motor.</li>↵
11   ······<li>Do·not·overoil·the·valve.</li>↵
12   ······<li>Rotate·the·shaft·about·its·axis.</li>↵
13   ······<li>Potlife·of·mix·is·approximately·4·hours.</li>↵
14   ······<li>We·advis
15   ······<li>A·maximu      STE Dictionary. The keyword 'advise' is not       i>↵
16   ······<li>Oil·the·       approved. Alternatives: TELL (v), RECOMMEND
17   ····</ol>↵                (v). Possible replacements: "tell", "recommend".
18   ··</body>↵                tell
19   </html>↵                  recommend
20   ◆
```

Figure 2: An XML editor integrated with an ASD-STE100 checker

Most 'structured writing' is only partly-structured writing. XML editors validate the high-level structures such as headings and paragraphs. But, the words that a technical communicator can use are not specified. Although the XML editor that is shown in Figure 2 finds terms that are not approved in ASD-STE100, it does not prevent the technical communicator from using those terms.

Usually, structured writing does not specify the permitted sentence structures. For example, the following sentences have the same meaning:

- Give the accident report to your manager.
- Give your manager the accident report.

Bischoff, Picart, and Rames (2014) agree. They write, "Our experience has shown that software like XML editors and languages like DocBook and DITA only provide an outline, a global structure, to technical writing. Individual authors fill pages with their own words and teams need overarching material to refer to in order to achieve stylistic and grammatical harmony."

For a document to be fully structured, the permitted terms and the permitted sentence structures must be specified. And a validating XML editor must give a validation warning if a technical communicator uses other terms or sentence structures. ASD-STE100 helps to give structure at the sentence level. Refer to 'Structure and discipline at the sentence level' on page 44.

Projections for the future

Many of the changes in technical communication are driven by developing technology. These changes are happening quickly and in ways that are difficult to predict. Nevertheless, in these recent articles the authors have tried to predict the future. For example, Perlin (2016), in predicting the impact of mobile technology, writes that technical communicators should:

> Refocus our writing: look at mobile first...Write to the extreme standard of mobile to improve and extend all our writing.

> Write shorter, simpler, more focused material...Consider moving away from full sentences to sentence fragments.

ASD-STE100 supports writing shorter and more focused text but it does not permit writing in fragments because fragments can create ambiguity.

Urbina (2014) explores the effects of augmented reality, wearable technology and the Internet of Things. On writing, Urbina says that wearable technology will make writing more complex:

> Technical communicators will have to write not just one deliverable, but create content that is designed to be accessed during one task across various devices...Imagine an engineer reading quick steps on an on-device screen, with diagrams or reference tablets stepping along in sequence on their tablet.

On augmented reality, Urbina suggests that organizations will be making apps instead of manuals and relying more on videos and graphics. In this case, where text is little or not used, ASD-STE100 will not be relevant.

In our experience, organizations are at different stages along the path to adopting these technologies. Many organizations still rely on written documentation, especially in safety-critical industries. Until this changes, ASD-STE100 still has a vital role to play in today's fast-moving world.

Conclusion

Even with the future trends that we have mentioned, ASD-STE100 will remain an important tool for technical communication in the 21st century. We will leave the final word to Orlando Chiarello, Chair of the STEMG, who was interviewed in 2014. He said (Leszinsky, 2014):

> Today, the success of Simplified Technical English is such that it is now rapidly moving outside its intended domain of aerospace...there is a worldwide growing interest towards Simplified Technical English, including...the Academic world and language professionals.

References

Almqvist I. and Hein A. S., (1995). Defining ScaniaSwedish – a Controlled Language for Truck Maintenance. In *CLAW 96, Proceedings of the First International Workshop on Controlled Language Applications*, pp. 159–164. KU Leuven, Belgium, 1996. Available at: http://stp.lingfil.uu.se/corpora/scania/ash961.html [Accessed 8 Feb 2017].

ASD (2012). *S1000D International specification for technical publications using a common source database*, Brussels, ASD.

ASD (2017). *ASD-STE100 Simplified Technical English: International specification for the preparation of technical documentation in a controlled language*, Brussels, ASD.

Avrahami H. (2016). *Delivering Consistent Content Experiences Across Digital Touchpoints.* Available at: http://thecontentwrangler.com/2016/07/11/delivering-consistent-content-experiences/ [Accessed 8 Feb 2017].

Baker M. (2015). *What is Structured Writing?* Available at: http://techwhirl.com/what-is-structured-writing/ [Accessed 8 Feb 2017].

Betts R. (2006). Wycliffe Associates' EasyEnglish. *Communicator*, 2005 (Spring), pp. 28–31. Peterborough, ISTC. Available at: www.easyenglish.info/about-us/articles/communicator.htm [Accessed 8 Feb 2017].

Bischoff F., Picart E. and Rames C. (2014). *Tools and Methods to Achieve Consistency in Technical Documentation,* Research Report, Université Paris Diderot. Available at: https://f.hypotheses.org/wp-content/blogs.dir/1236/files/2015/11/2014-Bischoff_Picart_Rames.pdf [Accessed 8 Feb 2017].

Braster B. (2008). Controlled language spreads its wings. *Communicator*, 2008 (Winter), pp. 33–34.

Doherty P. and Chiarello O. (2014). *Human Error: Controlled Language in Operating and Maintenance Manuals supplied to Ships.* Available at: https://www.sailors-club.net/maritime-files/file/3291-human-error-controlled-language-in-operating-and-maintenance-manuals-supplied-to-ships [Accessed 12 Apr 2017].

Eveland J. S. (1990). A search for clarity in technical documents. *Retrospective theses and dissertations.* Paper 160. Available at: http://lib.dr.iastate.edu/cgi/viewcontent.cgi?article=1157&context=rtd [Accessed 8 Feb 2017].

FAA (2014). *The Operator's Manual for Human Factors in Aviation Maintenance.* Available at: www.faa.gov/about/initiatives/maintenance_hf/library/documents/media/human_factors_maintenance/hf_ops_manual_2014.pdf [Accessed 8 Feb 2017].

Hackos J. (2012). *Minimalism updated 2012.* Available at: www.infomanagementcenter.com/publications/e-newsletter/december-2012/minimalism-updated-2012 [Accessed 8 Feb 2017].

ISO (2015). ISO/TS 24620-1:2015. *Language Resource Management – Controlled natural language – Part 1: Basic concepts and principles.* Geneva, ISO.

Information Mapping (2016). Available at: www.informationmapping.com/en/structured-writing [Accessed 8 Feb 2017].

Jackson M., Crouch S., Baxter R. (2014). *Software Evaluation Guide.* Software Sustainability Institute. Available at: www.software.ac.uk/software-evaluation-guide [Accessed 8 Feb 2017].

Kaiser H. (2016). *Simplified Technical English – a globally proven trendsetter.* Available at: www.tcworld.info/rss/article/simplified-technical-english-an-internationally-proven-trendsetter/ [Accessed 8 Feb 2017].

Kohl J. R. (2008). *The English Style Guide: Writing Clear, Translatable Documentation for a Global Market.* SAS Institute, Cary, North Carolina.

Kuhn T. (2013). *A Survey and Classification of Controlled Natural Languages.* Available at: http://attempto.ifi.uzh.ch/site/pubs/papers/kuhn2013cl.pdf [Accessed 8 Feb 2017].

Leszinsky L. (2014). *Meet the speaker: Orlando Chiarello, introducing ASD-STE100.* Available at: https://prozcomblog.com/tag/asd-ste100/.

Muegge, U. (2008). *Rules for machine translation.* Available at: www.muegge.cc/controlled-language.htm [Accessed 8 Feb 2017].

Muegge, U. (2009). Controlled language – does my company need it? Available at: www.tcworld.info/e-magazine/content-strategies/article/controlled-language-does-my-company-need-it [Accessed 8 Feb 2017].

Nerrière J.-P., Hon D. (2009). *Globish the world over*, International Globish Institute.

Ogden C. K. (1940). *Basic English: A General Introduction with Rules and Grammar.* Kegan Paul, Trench, Trubner & Co, London.

PEC (2016a). Plain English Campaign. Available at: www.plainenglish.co.uk/about-us.html [Accessed 8 Feb 2017].

PEC (2016b). *How to write in plain English.* Plain English Campaign. Available at: www.plainenglish.co.uk/how-to-write-in-plain-english.html [Accessed 8 Feb 2017].

Perlin N. (2016). How mobile is changing our industry. *Communicator*, 2016 (Summer), pp. 40–44.

Remedios R. C. (2003). *A Specification and Validating Parser for Simplified Technical Spanish*, MSc thesis, University of Limerick. Available at: www.csis.ul.ie/staff/Richard.Sutcliffe/ruiz_cascales_thesis03.pdf [Accessed 8 Feb 2017].

Rolls-Royce (2013). *Common Technical Data.* Available at: www.slideshare.net/TCUK_Conference/tcuk-2013-rollroyce-common [Accessed 8 Feb 2017].

SEC (1998). *A Plain English Handbook: How to create clear SEC disclosure documents.* Washington, U.S. Securities and Exchange Commission. Available at: www.sec.gov/pdf/handbook.pdf [Accessed 8 Feb 2017].

STEMG (2016a). Simplified Technical English Maintenance Group. Brussels. Available at: www.asd-ste100.org [Accessed 8 Feb 2017].

STEMG (2016b). *Service information letter / Position paper.* Simplified Technical English Maintenance Group. Available at: www.asd-ste100.org/service-letter.html [Accessed 8 Feb 2017].

Thrush E. A. (2001). Plain English? A Study of Plain English Vocabulary and International Audiences, *Technical Communication* 48(3), pp 289–296.

Urbina, N. (2014). Surviving accelerating change. *Communicator*, 2014 (Summer), pp. 25–29.

3

Why do we take the people out of our writing?

Kirstie Edwards

Abstract   What and how we write conveys a representation of ourselves, and we have many choices over how we construct this 'self-representation'. The choices we make over what to include and how conveys messages about us. In academic and professional writing we work within many constraints and often leave ourselves out as much as possible. I argue in this chapter that we should aim to bring more of ourselves into our writing and keep the 'constructed' identity in our writing as close to our genuine personalities as possible. I'm here writing this for you to read.

Keywords   constructed identity; personal voice; self-representation; self-revelation; social interactive writing

Introduction

You are probably already judging me and what I've written as you read these words. I'm making decisions about word choices, phrases, content and my views, while you're looking at my overall style and deciding whether I'm OK or not. According to the ancient Greek philosopher, Aristotle, a speaker's character is (almost) the most effective means of persuasion they possess (Brahnam, 2009), so I'm trying to keep me in this writing, to have the best chance of persuading you of my views.

Voice in writing can be defined as the communication of individual presence behind the written words (Narayan, 2012, cited from Flowerdew and Wang, 2015). Stock and Eik-Nes (2016) in their review of research into 'voice' in academic texts concluded that "there is no voiceless text" (2016, p. 10). Ivanič and Camps (2001) argue that writers through their choices of linguistic and other resources project a voice, and, "There is no such thing as 'impersonal' writing, because writers convey messages about themselves," (Ivanič and Camps, 2001, p. 4).

On the other hand, Hyland and Jiang (2017, p. 41) write in their discussion of the meaning of informality in academic writing: "The conventions of formality mean that, as far as possible, authors leave their personalities at the door when they sit down to write." So we have researchers arguing we are in the text and conversely writing that traditions (imposed through education, institutional guidelines, professional practice or other means of transferring norms or ways we should do things in certain contexts) restrict how much of our personalities we put into our writing.

Some researchers (Ivanič and Camps, 2001) claim that all sorts of decisions in writing construct the identity of the writer so that writing always conveys a self-representation. Although conformity to the norms of a certain type of writing, for a certain purpose, in a certain context may be moulded by social norms, writers still have the choice of conforming or resisting these pressures. So at one end of the spectrum, we could write just as we speak, freely and without any restrictions, in which case we would be representing our true personalities, as for example we might in an email, and at the other end of the spectrum, we write conforming to restrictions of a certain community or for a certain type of writing and create a social identity, perhaps still portraying some of ourselves, but also hiding much of ourselves. "Writers often convey, through the varied linguistic and other resources they draw upon in their writing, a multi-faceted self," (Flowerdew and Wang, 2015, p. 83). So there might be a natural me in this writing or there might be a socially constructed me who has conformed to the style guide written for this book and other social norms of writing for this type of book, readership and other constraints. Or I may have completely fabricated myself in this chapter and created a persona to help persuade you of my viewpoint.

My overall message is that rather than remove ourselves from our writing, it might help both writers and readers in the writer–reader interaction, if we left ourselves right in there. It might avoid the block which inexperienced writers feel when first socializing into (or learning the ropes of) scientific and technical or any type of non-fiction writing, it might help simplify writing and thereby expand the audience able to understand it and it might help readers to evaluate the truthfulness of facts and knowledge reported and evaluate the ideas we convey, based on not only the content, but also on our identity – who we are and our credibility.

Based on a literature review and on the premise of writing as a social interaction (Nystrand, 1989; Hyland, 2005), I will revisit the debate about whether scientific and technical writing style should be more personal. In 1989, Turk and Kirkman wrote about the "accepted dogma of scientific writing… [with] no references to the person doing the work," and argued: "It is artificial to avoid personal references in scientific writing," (1989, p. 112). Twenty-eight years later, I observe within higher education (HE) and from most scientific and technical publications, which I read, that an impersonal style is still often preferred. Great strides have been made to improve clarity of writing with the Plain English Campaign (2016), but within some HE institutions, it seems we are still teaching 'objective' and impersonal writing styles.

There is a strong link between the practices in academic institutions and the practices in professional disciplines. Education, enculturation/ socialization or indoctrination (and notice that I've used a negative, not neutral, word with 'indoctrination') of young writers into the writing of their disciplines starts in education, where they are taught the norms of academic, scholarly, scientific and technical writing. Whether these young students become engineers, scientists, doctors, nurses or technical writers, this foundation to their writing styles may persist. Kirkman (1975, p. 200) writes "…the way in which we encourage young writers to think about choice of verbs and constructions is of outstanding significance, because it goes a long way towards establishing their overall 'mental set'." My perspective in this chapter therefore focuses a great deal on academic writing as the (educational) starting point for other professional writing.

This chapter dips into many disciplines to connect some ideas and trends related to how much of ourselves we should bring into our writing. Within the word count, therefore, I haven't been able to provide details on each research study mentioned or genre of writing referred to. The result is therefore naturally a rather diffuse network of connections, which I hope will help clarify the directions of my thoughts and arguments on the topic.

First I explain some of the problems which might (in part) be addressed by a more transparent style of writing, with us in our writing more. Then I touch on three theories which frame different ways we can view how we represent ourselves in our writing. Having identified different ways we can do this, I explore example textual strategies and what research is telling us about their use. As a corollary to removing ourselves from our writing, I then introduce some research on bringing personas into writing. Finally, I summarize my projections for the future and conclude on the topic.

What is the problem? Arguments and trends in practice

In academic, technical and scientific writing we are dealing with facts, knowledge and truth. I was traditionally taught that technical writing should show no bias, opinion or emotion; in fact, I was trained to remove myself from my writing.

In education today, the focus has shifted from rote learning to developing skills in finding relevant information and evaluating its quality. The trend of increasing information overload emphasizes the necessity to evaluate the quality of what we are reading accurately. We don't have time to read everything, so we need to filter out only the most important and credible information to read efficiently and find what we need. There is an increasing need for transparency and openness ('open bias') not just in conflicts of interest or other causes for impartiality, but also in exactly who did what and why or what authors believe and why. Readers need to be able to evaluate the writers and their writing and decide whether the information they are reading is likely to be true.

In today's world of 140-character tweets and text messaging, young people are reading lengthy texts less and less. Whereas academic writing has been historically hard for new students to understand, for today's young students, it's like "translating a foreign language" (Stewart, 2010, p. 749). Stewart asks in a letter to *Science Magazine* (2010, p. 749), "if academic writing were to become less formal and less terse, would the communication of scientific ideas suffer?" Others take the view that specialist writing has no need to be understood by a general audience: "Academic writing is highly specialized: most people do not read it… where is the need for it to match more popular registers of communication in terms of style?" (Minton, 2015, p. 3).

I argue that Minton's (2015) perspective is flawed: I am a proponent of Plain English, minimalism and avoidance of jargon. It may be true that most people do not read academic writing, but for me that is an argument for matching it more to the popular communications in terms of style to encourage and facilitate its reading. Academic style serves as a deterrent or block to many undergraduates and non-experts generally.

Many non-experts seek to understand scholarly and specialist writing, which is publicly available online, related to health conditions, for example. Minton (2015) assumes that because the knowledge originates from a specialist, there is no need for it to be communicated in a more generally understandable style. Readers of healthcare information online may not be medical experts, but they are readers, who (I argue) should not be excluded from being able to interpret knowledge and ideas, because they are not experts. On the contrary, the fact that the information is accessible to non-experts, strengthens the argument that it should be understandable in terms of meaning and transparent in terms of the personality writing it.

With globalization, English has become a global language for scientific communication and users of English as an additional language (EAL) now outnumber those who speak English as their first language (Flowerdew and Wang, 2015). EAL novice scholars already disadvantaged in their level of English are therefore also faced with the difficulty of familiarizing themselves with practices of the scientific writing community.

In my work as a tutor for students with a learning difference, I repeatedly find that although students balk at writing down their ideas based on what they have learnt, they articulate them quite comfortably in conversation. Between their conversation and the written word lies a huge barrier of fear over formulating adequate 'academic language'. Often their attempts at academic writing appear to portray a persona other than the student I know, a persona which sounds pompous and restrained and distant. As Stewart (2010) argues, would simple, honest language really take the science out of science? And would it be less credible if the author was less hidden?

However, some argue in the opposite direction. For example, Budgell (2013) claims that biomedical English differs so much from general English, that it should be called 'biomedical language'. He argues that "the average native English speaker has little advantage over the average, for example, native Chinese speaker when trying to become fluent in the languages of biomedicine and health. In other words, biomedical language is a second language for all of us" (Budgell, 2013, p. 229). This is a particularly interesting stance, because Hyland and Jiang (2017) have just completed some research exploring the trend towards informal writing, which their data suggests was highest in the closely related discipline of biology.

Finally, in education, students repeatedly ask me in writing tutorials, "How can I conclude on the topic without giving my opinion away?" They are requested to argue a point in an assignment in a balanced way, but to disguise their own stance on the matter. My response is always that the whole point of the assignment is to persuade the reader of their opinion or view, but I can understand why students are baffled by this when they are constantly warned not to write about their own opinions. They are taught to hide their personal views, but somehow deliver them.

To summarize, there's more information out there, we need to scan it quickly and evaluate accurately what's worth reading, which may be an easier task if it's understandable to everyone and written honestly and openly. Three aspects of writing can contribute to this: clear writing in plain English, strong summarising skills (so that people can read a short text and decide whether to read the entire text) and self-representation as genuine as possible, unrestricted by norms of writing in certain disciplines or communities. This chapter focuses on self-representation.

Relevant theory

I visualize written communication framed by Nystrand's (1989, p. 73) social interactive model of writing, which identifies "text as a communicative event" with a context of production and reception. I've written in one context, anticipating your understanding (and how to optimize that happening) and you are reading in another context. My representation of the meaning (in this chapter) is designed to reach your mind and hopefully a common ground of understanding, a meeting of minds. It's a social process because you are already involved together with the ISTC writing community, peer reviewers and – to a certain extent – other editors and publishers of similar works in the past, who have developed norms of practice for how I should write this. So the first point I want to make under my 'theory' heading is that writing is a social interaction, with many potential constraints. The question I address in this chapter is: do I need to remove me from the writing to achieve the meeting of minds, or can I deliver the message in writing in a way that not only conveys my meaning but also conveys a more genuine me, allowing more accurate evaluation of me, and closely related to that a more accurate evaluation of the proposition I'm making in my writing?

Hyland (2005) also argues that writing is a social interaction and that we cannot write in a social vacuum. Instead, he argues, we cannot remove ourselves from the text; we manifest ourselves either explicitly or as metadiscourse using linguistic strategies. Hyland's (2005; 2015) (metadiscoursal) linguistic strategies include hedges (possible, might, perhaps), boosters (clearly, obviously, surely), attitude markers (agree, prefer, hopeful, remarkable) and self-mention (first-person pronouns and possessive adjectives). Even who and how citations are integrated in academic writing can differ by identity (see Flowerdew and Wang's 2015 review on this).

Ivanič and Camps (2001) also argue that many choices in writing relating to content and form and even presentation convey messages related to self-representation. Burgess and Ivanič (2010) discuss four aspects of identity in writing: social, autobiographical, discoursal and authorial. Social identity is framed by the norms of the discourse community (particular writing community or discipline, for example medical researchers writing in scholarly journals). Autobiographical identity relates to life experiences, values, beliefs and social positioning of the author. I, for example, am writing this based on my personal (and therefore unique) life experiences and what I've learnt in the world up until this point. Discoursal identity relates to representation of self through writing choices (some examples of which I discuss below) and the authorial self relates to how authoritative the author feels and how strongly they assert their position. With Ivanič's model, there are thus many factors influencing how I portray me to you through this writing.

Systemic-Functional Linguistics (SFL) is a theory of language which focuses on what language does and how. Halliday et al. (2014) identify three macro-functions of language: ideational positioning, interpersonal positioning and textual positioning. Ideational positioning relates to the ideas, concepts and views we have, interpersonal positioning relates to different degrees of self-assurance and certainty and different power relationships between the writer and the reader and textual positioning relates to the different views of how a written text should be constructed.

That presentation in writing conveys something of ourselves has also been argued (Sless, 2004; Nord and Tanner, 1993). Carelessly presented documents show little respect to readers, whereas carefully presented documents are more likely to create a good impression; they suggest that the same care has gone into writing the text as in its design. In the same way that our personal presentation at an interview makes a first impression, so too does the first visual image of a text or document. That first image of a written text is conveying something about us and the effort we have made for you the reader, so in part the way I've presented this chapter conveys something about me, which you can judge.

In summary, written communication is a social interaction, which includes content (ideas), relationships (relative authority and power) and choices of textual or linguistic strategies. Choices we make in how we convey ourselves through the content and form of our writing, its metadiscourse and presentation (intentionally or unintentionally) determine what the reader perceives as the author's identity. My argument in this chapter is that the constructed identities in writing should be closer to genuine author identities than they are in current academic, scientific and technical writing. To explore this idea further, I now discuss two textual choices, which reflect markers of self-representation in writing: use of active voice and use of first person pronouns.

Textual positioning

The active voice is a good starting point for trying to bring ourselves into our writing, because it emphasizes the agent (who is doing) of an action, in this case me as the writer. I started this chapter believing that there was still a prevalence of passive writing in academic and scientific publications. To a certain extent, I was wrong. A preliminary literature search identified that many journals now indicate in their instructions to authors either a preference for the active voice or at least recommend use of the active voice where possible (Every, 2015). Historically, in academic and scientific writing there appear to have been shifts both towards and away from detached writing styles. Ding (1998, cited in Leong, 2014) analyzed texts from the 18th century to the end of the 19th century and found that early scientific writing favoured the active voice; there was then a shift towards passive voice in the 20th century to meet the increasing demand for objectivity and the need to represent the world "in terms of objects, things and materials rather than humans" (Ding, 1998, cited in Leong, 2014, p. 1).

While there are some circumstances when the passive is appropriate, Rude (2006) argues that the active voice brings energy and establishes responsibility in writing. Barker (1998, p. 331) in his book titled *Writing software documentation* points out how passive sentences fill up with nouns and "Nouns clutter things up. Verbs get things done." The passive voice has been critiqued as vague (possibly obscuring identities and bias), lengthy, more complex to use and pompous – all critiques which Minton, (2015) challenges in his paper. The active voice in comparison, is assumed to be more concise, informative, easier to use and less pompous. However, Hyland (2001) recognizes an underlying assumption related to persuasive academic writing, that objectivity and removal of self from writing represents a humble and self-effacing approach, best suited to gaining acceptance of claims.

Ding (2002) argues that use of the passive voice in scientific writing is closely linked to two social values, falsifiability in science and the cooperation among working scientists. Falsifiability relates to being able to repeat work to test and verify findings and cooperation refers to how knowledge is developed collaboratively by scientists in a particular field through time. Ding argues that these two social values of scientific study are expressed through the passive voice with "personal reservations... removed from what is being performed, and personal qualifications and personal privileges ...taken away from what is being observed" (Ding, 2002, p. 146). Further, Ding (2002) argues that we build on communal experience in science through things being acted on (and described using the passive voice). Ding claims that: "The passive voice in scientific writing conforms to the thing-centred scientific work, it establishes a common domain for scientists to work in" (2002, p. 152). This seems in direct conflict with Hyland's (2002; 2016) argument that we cannot remove representations of ourselves from our writing (entirely) and also an argument by Baratta (2009) that the passive voice can be used to emphasize stance. Self-representation in writing integrates identity in a multitude of ways, not simply through a dichotomous decision to use active or passive voice, but rather through many decision-making processes, as discussed in the theory section above.

Today, many scholarly and scientific journals allow, recommend or even prefer the active voice and specify this in their author guidelines (Every, 2015) encouraging transparency and accountability in writing. Leong (2014) refers to the clear shift towards advocating active voice in writing style guidelines. Leong (2014) claims that use of the passive may remain stable at around 30% frequency in scientific writing, but that education is required so that writers know that they don't need to stick to the passive.

However, guidelines are not the only influence on writing style; there are many influencing factors in the journey to publication. "People are constrained, but not determined, by the dominant disciplinary, professional, gender and political identities which are set up by the conventions of specific genres and the practices which surround any act of writing. We all bring multiple possibilities to any act of writing which carry the potential to challenge the pressures to conform to dominant identities" (Hyland, 2016, p. 46–7). Influencers in the whole process start from the author and their own level of enculturation in their discourse community, through to organizational style guides, peer reviewers, editors and publication style guides.

In spite of guidelines advocating the use of active voice in writing to move away from distanced, objective writing, my experiences in HE (at four universities in Europe) suggest that traditionalist views about the need for passive writing still exist. Leong (2014) reports a mini informal survey of 99 students, 90 of whom felt that the passive voice should be used as the norm in academic writing. I recently learnt from a student that her lecturer did not permit use of the first person in a piece of reflective writing. (Reflective writing prompts students to think about an experience and relevant knowledge and theory, make sense of what happened in the light of available knowledge and identify how their response to the situation might be improved on another time. For example, a nursing student might reflect on an experience with a patient in the light of a new best-practice guideline she had learnt about.) My feeling was how could you take the 'me' out of an internal reflection? What advantage is there of distancing yourself in the reporting of your own internal voice? Lea and Street (1998, cited in Logan, 2012) found that even within the same courses, individual tutors had different expectations about when use of the first person was appropriate.

The above-mentioned anecdotal evidence of a tendency to adopt the passive is also corroborated by research. Millar et al., (2013) analyzed the frequency of passive constructions in the research reports of randomized control trials (RCTs). RCTs are the gold standard in medical research and so were considered an informative type of writing to explore. For reference against 'general English', articles from *The New York Times* were also analyzed. Millar et al., (2013) found that style guides discouraging or not discouraging use of the passive had a significant effect on use of the passive voice in articles published in them. (This might seem obvious, if editors have influence over the guidelines and also over approval for publication of articles, but the picture may be more complex than it first appears; read on). Passive constructions were used twice as often in RCT articles than in the 'general English' texts. So there was a considerable difference between the two sorts of texts. Reporting everyday activities in a newspaper did not appear to require the objective distancing expected of the medical trial reports. Related to academic writing, Hiltunen (2016) found variations in frequencies of use of the passive when researching essays in different disciplines. Passives were more frequent in physics and medicine essays than in law and literary criticism essays. Use of the passive thus seems to vary by both discipline and target audience or genre.

Amdur et al. (2010) analyzed 90 articles of three types from three journals. They compared frequency of the use of passive voice measured by percentage of sentences using the passive and compared the medical articles with articles from *The Wall Street Journal*. They found 20–26% passives in the medical articles, compared with 3% in the newspaper articles, and argue therefore that this is evidence of overuse of passive in medical journals. These researchers raise the question of whether it is publishers who are to blame for this, not writers, which is interesting in the light of the trend for journals to advocate active voice and Millar et al.'s (2013) research suggesting that such guidelines tend to have influence.

Research will always have its flaws, of course; for example, measures vary in the research I've cited here. Additionally, all these studies focus on articles which passed the gatekeepers' influences so that they are not a fair representation of writing by medical researchers (or newspaper reporters) but rather a representation of the articles which meet the criteria for approval to be published. However, setting these limitations aside, it seems that while guidelines are advocating active voice in scholarly publications, passive is still being used, although there is an increasing need for transparency and clear communication for evaluative purposes in the ever-growing mass of communicated knowledge and ideas.

Recommendations from the literature I've read vary slightly, but I summarize them here. Elliston (2008, p. 31) writes in the Journal of the European Medical Writers Association: *The Write Stuff*, that "the active voice is now preferred unless otherwise stipulated. It produces clearer, more direct language… Some writers feel that the passive voice sounds more modest and 'scientific', but the foremost aim should be for clarity and directness, avoiding superfluous words." She continues, however, to give examples of where use of the passive is appropriate. Leong (2014) also recommends that the active is not always appropriate. Overuse of the passive should be avoided rather than advising complete avoidance, whereas Amdur et al., (2010) advise that journals should specifically recommend the use of less than 10% passive voice. Millar et al. (2013, p. 411) recommend formative style guides: "Advice that authors should use active voice whenever possible is over-simplistic." They argue that this advice implies that the active and passive are interchangeable, which they are not. These researchers conclude that style guidelines should reflect the complexities of actual use of the passive and active voice (see Elliston, 2008 for some examples of appropriate uses). Jutel (2007) also recommends decision-making guidance in style guides, rather than advice to use the active or passive voice.

Minton (2015) describes the passive voice as an intrinsic part of the English language, to be used appropriately to maintain the natural flow of writing and argues against generalizations that passive voice is ambiguous or longer. He also recommends that guidelines should ask for appropriate use rather than be prescriptive in dictating either active or passive use. He advocates first writing from the viewpoint of given–new and that the choice of appropriate active or passive voice will then be clear. "Keeping the topic of discourse in subject position and presenting new information later in each sentence is the natural pattern in English and will normally determine the selection of voice automatically" (Minton, 2015, p. 9). Similarly to Millar et al. (2013), Minton (2015, p. 8) points out that the active and passive are not interchangeable and that "Brevity is no advantage if it comes at the expense of clarity and natural word order." He argues that context is also important to meaning (p. 9) and can convey the identity of whoever is 'doing' by assumption.

Use of the first person is another textual strategy for bringing ourselves into our writing. While Einstein recommended: "when a man is talking about scientific subjects, the little word 'I' should play no part in his expositions" (cited from Hyland, 2001, p. 208), Kirkman (1975, p. 198) points out that experts (such as scientists) can quite comfortably speak of their work using first person pronouns so that he struggles to understand why "the personal active phrasing acceptable in serious discussion is not acceptable in serious writing." Following feedback from tutors that Logan (2012 p. 782), "needed to own the work more" and use less of the passive, she analyzed her own writing, exploring frequency of first person use. Logan (2012) concluded that: "Uses of the passive were to distance myself and to be less specific." She writes, "I was trying to lend some kind of authority to [my] this early writing" (Logan, 2012, p. 782). Jutel (2007) points out that nursing students are traditionally trained to keep themselves out of their note writing in health practice. "Revealing our identity was a legal responsibility, but hiding it in our notes was a duty" (Jutel, 2007, p. 3). There thus still appears to be some confusion over whether we should or shouldn't be in our writing both while learning about writing and in professional writing contexts.

Bringing ourselves into our research is increasingly advocated in qualitative research (Jasper, 2005). "Writing in the first person acknowledges the centrality of the writer, writing reflexively cultivates a self-awareness" (Jasper, 2005, p. 250). Jasper (2005) continues to argue that such dialogue promotes an analysis and understanding of important issues in research. By reporting our thoughts, feelings and decisions as we research, we are exposing ourselves and avoiding bias. "Reflective writing acknowledges the subjective nature of the researcher's interaction and interpretation of the data, providing the decision-trail within the public domain… and transparency of the processes leading to conclusions being presented" (Jasper, 2005, p. 250). Plainly, reflective writing brings us into our writing and would be harder to do without use of the first person.

Hyland and Jiang (2017) from their study of a corpus of 2.2 million words from leading journals in four disciplines at three periods showed that there has only been a small increase in linguistic features representing informality. The increases were mainly in the hard sciences, whereas social sciences had become slightly more formal. Of particular note was the 213% increase in use of the first person in the biology journal articles analyzed, from 1965 to 2015. Hyland and Jiang (2017, p. 48) researched other representations of informality than first-person pronouns and warn from the results that "any trend towards informality is not cut and dried" (and notice the idiom here). Indeed, in an earlier study, Gillaerts and Van de Velde (2010, cited in Flowerdew and Wang, 2015) concluded that use of linguistic markers of interpersonality had diminished over the 30-year span of the corpus they studied.

Rudolf Flesch, readability expert, writes: "if you want to write like a professional, you have to get used to the first-person singular. Never mind the superstitious notion that it's immodest to do so… If you want to write well, about anything at all, you must be prepared to face the consequences and portray yourself quite mercilessly whenever the occasions arises" (cited from Amdur et al., 2010, p. 103).

Hyland also argues that "Self-mention is important because it plays a crucial role in mediating the relationship between writers' arguments and their discourse communities, allowing writers to create an identity as both disciplinary servant and persuasive originator… Their intrusion helps to strengthen both their credibility and their role in the research, and to help them gain acceptance and credit for their claims" (Hyland 2001, p. 223); "…controlling the level of personality in a text becomes central to building a convincing argument" (Hyland, 2005, p. 173).

Bringing non-people into our writing!

In instructional writing, there appears to be a trend towards bringing people into writing. Van der Meij (2009) discusses the move from around the year 2000, towards managing emotions in user guides showing empathy and documents being attractive, motivating and fun to read, like the *For Dummies* books. These books aim to give instruction, but also to reduce frustration, boredom and insecurity, which are often experienced when learning how to use software (Van der Meij, 2009). To do this, the books need motivational and affective elements: "They make the author's presence obvious" (Van der Meij, 2009, p. 280). The *For Dummies* series is highly successful with fun and humorous approaches to information design; 200 million copies of more than 1,400 titles have been sold since 1991 (Willerton and Hereford, 2011).

Coney and Chatfield (1996) analyzed one of the first *For Dummies* books and claimed that the presence of a clear authorial voice was one of the most important reasons the books were so successful. They analyzed *Microsoft Word User's Guide* and found no individual mentioned, whereas in the *WordPerfect 6 For Dummies* guide, the author is introduced in detail. The author (Dan Gookin) is separate from the product (not an employee of the product's manufacturer), so that he can critique it with honesty. The author provides humour and entertains; he brings trust and a social interaction with the reader. He's right there in the writing.

There also appears (oddly) to be a movement towards introducing false personas to whom readers can relate, into texts. This seems absurd to me; we are writing on the premise of removing ourselves to remain objective, but including a third party, who doesn't exist. I will now discuss some research using these personas or false agents and then relate this research to my discussion above.

Van der Meij (2008) introduced co-user characters into manuals to test whether they improved user confidence. There appeared to be no difference in the user confidence, which may have been due to the manuals having been optimized anyway (even without the addition of the co-users) or to the co-users being a distraction to learning. In another study, Van der Meij (2013) studied the use of motivating agents in printed software tutorials. His sample comprised 49 students using printed manuals with or without the agents and he measured skills and self-efficacy before and after training. The agent condition scored higher on skills and self-efficacy after the training. A study by Loorbach et al., (2013) found that reader confidence in learning was improved through the use of personal stories, again using matching personas, which readers related to.

Engagement and the reader–writer relationship thus seem strengthened by both fore grounded authors (as in the *For Dummies* series) and by false personas as in the research cited here. These seem strong arguments to me for introducing ourselves more transparently in our writing to improve reader engagement and the reader–writer relationship.

Projections for the future

I anticipate that bringing constructed identities (through representations in writing) closer to real identities, for transparency of identity and reading of credibility will become more valuable as knowledge expands and we become more accustomed to reading shorter texts relating to multiple disciplines. Hyland and Jiang (2017) recognize a formal/informal continuum from legal documentation to email conversations. By formality they refer to the use of technical and abstract vocabulary, complex sentence structures and impersonal voice. The risk of informality is ambiguity, misunderstandings, lack of objectivity and possibly a failure to work in cohesion with a body or community of practice as Ding (2002) refers to. But Hyland and Jiang (2017) point out that in some fields, scholars are making writing choices to engage readers and form personal connections with them.

Conclusion

Written communication is a social interaction, which includes content (ideas), relationships (relative authority and power) and choices of textual or linguistic strategies. Choices we make in how we convey ourselves through the content and form of our writing, its metadiscourse and presentation determine what the reader perceives as the author's identity. My argument in this chapter is that the constructed identities in writing should be closer to genuine author identities than they are in current academic, scientific and technical writing.

The success of the *For Dummies* books has been in part attributed to the strong authorial voice, and some research on personas in instructional writing suggests these improve engagement with readers. This seems a further argument for leaving me in my writing. Two example textual strategies to achieve this are the active versus passive voice and use of first person. Interestingly, while style guides for many scholarly and scientific journals appear to be recommending active voice, and research suggests style guides have a significant impact on writing, other research suggests the trend towards informality is very small and slow. Researchers on the topic tend to recommend the active voice without being prescriptive; there are still occasions when the passive is more appropriate and this is one of the many decisions writers need to make.

Writing is all about decisions, some of which are constrained for sensible reasons, such as making sure we all know what is meant by the words 'pulmonary embolism', if we (or a group of doctors) need to do something about it. But some restrictions may be oppressive and prevent openness in writing. Keeping ourselves in our writing might help genuine reader–writer interaction, simplify writing, help people overcome barriers to writing and help readers to evaluate writing.

One final point; I'm not an expert in linguistics or identity and may have misunderstood concepts which I've read about, forgotten important points or unintentionally explained points poorly. However, I hope that I've prompted you to think about being in your writing. Thank you.

References

Amdur, R.J., Kirwan, J. and Morris, C.G. (2010). Use of the passive voice in medical journal articles. *American Medical Writers Association AMWA Journal*, 25(3), 98.

Baratta, A.M. (2009). Revealing stance through passive voice. *Journal of Pragmatics*, 41(7), 1406-1421.

Barker, T. T. (1998). *Writing Software Documentation: a Task-oriented Approach*. Boston: Allyn and Bacon.

Brahnam, S. (2009). Building character for artificial conversational agents: ethos, ethics, believability, and credibility. *PsychNology Journal*, 7(1), 9-47.

Budgell, B.S. (2013). The language of integrative medicine. *Journal of Integrative Medicine*, 11(3), 229-232.

Burgess, A. and Ivanič, R. (2010). Writing and being written: issues of identity across timescales. *Written Communication*, 27(2), 228-255.

Coney, M.B. and Chatfield, C.S. (1996). Rethinking the author-reader relationship in computer documentation. *ACM SIGDOC Journal of Computer Documentation*, 20(2), 23-29.

Ding, D.D. (2002). The passive voice and social values in science. *Journal of Technical Writing and Communication*, 32(2), 137-154.

Elliston, V. (2008). Tense and voice in medical writing. *The Write Stuff (Journal of the European Medical Writers Association)*. 17(1), 30-32.

Every, B. (2015). Biomedical editor: clear science writing: active voice or passive voice? Available at: http://www.biomedicaleditor.com/active-voice.html [Accessed 30 Nov 2016].

Flowerdew, J. and Wang, S.H. (2015). Identity in academic discourse. *Annual Review of Applied Linguistics*, 35, 81–99.

Halliday, M., Matthiessen, C.M. and Matthiessen, C. (2014). *An Introduction to Functional Grammar*. Routledge.

Hiltunen, T. (2016). Chapter 7: Passives in academic writing: comparing research articles and student essays across four disciplines. In López-Couso, M.J., Méndez-Naya, B., Núñez-Pertejo, P. and Palacios-Martínez, I.M. (Eds.), *Corpus Linguistics on the Move: Exploring and Understanding English through Corpora* (pp. 132–157). Leiden : Brill NV.

Hyland, K. (2001). Humble servants of the discipline? Self-mention in research articles. *English for Specific Purposes*, 20(3), 207–226.

Hyland, K. (2002). Authority and invisibility: authorial identity in academic writing. *Journal of Pragmatics*, 34(8), 1091–1112.

Hyland, K. (2005). Stance and engagement: a model of interaction in academic discourse. *Discourse Studies*, 7(2), 173–192.

Hyland, K. (2015). Metadiscourse. In Tracy, K., Ilie, C. and Sandel, T. (Eds.), *The International Encyclopedia of Language and Social Interaction*, 3 Volume Set (Vol. 1) (pp. 997–1009). John Wiley & Sons.

Hyland, K. (2016). *Teaching and Researching Writing*. 3rd Edition. New York: Routledge.

Hyland, K. and Jiang, F.K. (2017). Is academic writing becoming more informal? *English for Specific Purposes*, 45, 40–51.

Ivanič, R. and Camps, D. (2001). I am how I sound: Voice as self-representation in L2 writing. *Journal of Second Language Writing*, 10(1), 3–33.

Jasper, M.A. (2005). Using reflective writing within research. *Journal of Research in Nursing*, 10(3), 247–260.

Jutel, A. (2007). Actively passive: understanding voice in academic writing. *Nurse Author & Editor*, 17(2), 3.

Kirkman, J. (1975). That pernicious passive voice. *Physics in Technology*, 6(5), 197–200.

Leong, P.A. (2014). The passive voice in scientific writing. The current norm in science journals. *Journal of Science Communication*. 13(1), 1–16. Available at: https://jcom.sissa.it/archive/13/01/JCOM_1301_2014_A03 [Accessed 30 Nov 2016].

Logan, A. (2012). Improving personal voice in academic writing: an action inquiry using self-reflective practice. *Reflective Practice*, 13(6), 775–788.

Loorbach, N., Karreman, J. and Steehouder, M. (2013). Verification steps and personal stories in an instruction manual for seniors: effects on confidence, motivation, and usability. *IEEE Transactions on Professional Communication*, 56(4), 294–312.

Millar, N., Budgell, B. and Fuller, K. (2013). 'Use the active voice whenever possible': the impact of style guidelines in medical journals. *Applied Linguistics*, 34(4), 393–414.

Minton, T.D. (2015). In defense of the passive voice in medical writing. *The Keio Journal of Medicine*, 64(1), 1–10.

Nord, M. A. and Tanner, B. (1993). Design that delivers – formatting information for print and online documents. In Barnum, C. M. and Carliner, S. (Eds.), *Techniques for Technical Communicators* (pp. 219–251). Boston: Allyn and Bacon.

Nystrand, M. (1989). A social interactive model of writing. *Written Communication*, 6(1), 66–85.

Plain English Campaign (2016). Plain English Campaign. Available at: http://www.plainenglish.co.uk/ [Accessed 7 Dec 2016].

Rude, C.D. (2006). *Technical Editing*. 4th Edition. London: Pearson Longman.

Sless, D. (2004). Designing public documents. *Information Design Journal+ Document Design*, 12(1), 24–35. Available at: http://www.lyftingsmo.no/labelling/Presentations/David%20Sless,%20Designing%20Public%20Documents.pdf [Accessed 7 Dec 2016].

Stewart, M. (2010). Science education: flouting formality (Letter). 329(5993), 749–750. Available at: http://science.sciencemag.org/content/329/5993/749.3 [Accessed 7 Dec 2016].

Stock, I. and Eik-Nes, N.L. (2016). Voice features in academic texts – a review of empirical studies. *Journal of English for Academic Purposes*. Available at: http://dx.doi.org/10.1016/j.jeap.2015.12.006 [Accessed 30 Nov 2016].

Turk, C. and Kirkman, J. (1989), *Effective Writing. Improving Scientific, Technical and Business Communication*. London: E & FN Spon.

Van der Meij, H. (2008). Designing for user cognition and affect in software instructions. *Learning and Instruction*, 18(1), 18–29.

Van der Meij, H. (2013). Motivating agents in software tutorials. *Computers in Human Behavior*, 29(3), 845–857.

Van der Meij, H., Karreman, J. and Steehouder, M. (2009). Three decades of research and professional practice on printed software tutorials for novices. *Technical Communication*, 56(3), 265–292.

Willerton, R. and Hereford, M. (2011). Evaluating applications for an informal approach to information design: readers respond to three articles about nursing. *Journal of Technical Writing and Communication*, 41(1), 59–82.

4

A change in tone

Ellis Pratt

Abstract   Some organizations are changing the way they write user assistance, and are, for some content, not using the traditional, generally accepted, best practices. Through web analytics and other measures, they are reporting a noticeable benefit from making these changes. The reason for making these changes is due to people's relationship with technology changing, and a desire by organizations to deliver better forms of user assistance. However, there is no one right answer; you will need to experiment to see what works best for your organization.

Keywords   trends; tone and voice; software as a service; empathy

Introduction

If you were to be transported back 20 or 25 years, and found yourself in a classroom learning about technical writing, you'd probably find it was almost identical to classes on this subject offered today. Technical communicators tend to assume that the technical communication best practices that have been taught for the past 25 years, and even further back in time, are still the correct approach to take today.

Yet there are signs that are prompting some technical communicators to question whether the commonly accepted approaches are still appropriate in all situations.

Let me list a few examples:

- Mozilla, the developers of Firefox, reported a change to a more conversational and friendly style of writing help topics resulted in a 13.1% increase in page hits. The re-written pages were helpful to 800,000 more people per year, providing a significant reduction in support calls.

- Other leading web-based companies, such as Twitter and MailChimp, have help topics that most technical communicators would say were breaking the rules of technical communication best practice.

- At the tekom 2013 conference, Melanie Huxhold and Dr Axel Luther of SAP reported in their presentation that more users were viewing SAP's screencasts than reading the help pages (Huxhold and Luther, 2013). In other words, a medium where it is harder to search for specific information, and is more verbose, was more popular than online help.

- User documentation is often criticized for being boring and old-fashioned, and there is a commonly held belief outside of the technical writing community that 'no-one reads the manuals' – we create something of uncertain value.

We'll investigate whether the tried-and-tested writing methods from past decades still make sense today. We'll look at the reasons why some organizations are 'breaking the rules' with the user assistance they provide. We'll also look at the reasons for this change, the techniques used, and we'll explore the implications of this on the technical writing and localization teams.

Traditional writing principles

If you look at the timeline of technical communication standards, you'll see most of these emerged from the aerospace and mainframe computer industry sectors between 1960 and 1990.

The key milestones were:

- 1961 Quick Reader Comprehension (QRC)
- 1963 Hughes STOP – (Sequential Thematic Organization of Publications)
- 1967 Information Mapping
- 1974 SGML
- 1982 Information Types
- 1990 Minimalism and task-orientated instructions (Wikipedia, 2016).

We can see from this the writing style has essentially remained the same for more than 20 years. As for DITA itself, it is true to say it is a new standard: it was introduced in 2002 and approved as a standard in 2005. However, DITA's focus is on structuring and organising information around Information Types such as task, concept and reference. We would argue that DITA is a way to apply the minimalism style of writing.

These standards were written when technology could be described as big and scary for many users. When things went wrong, typically the consequences were expensive and sometimes dangerous.

Typically, technical communicators write content for a 'body of knowledge', courseware developers create training materials, and users generate their own content about how to master a product. Users get stuck, they seek out help in the body of knowledge, and hopefully resolve their problems.

These tend to be discrete 'islands of content': when users want to do more than just apply what they know, they tend to use the Internet-places such as YouTube and user forums – to learn, debate and make suggestions; technical communicators create online help, screencasts or printable user documentation; courseware developers create classroom training courses or elearning material.

This matches roughly with popular learning models (Bloom's Taxonomy of Learning and Kirkpatrick's Learning Evaluation Model), suggesting technical communicators are enabling users to understand and apply their knowledge (the conscious competence and conscious incompetence levels in the diagram below).

A focus on efficiency more than on quality

In recent years, a great deal of focus in technical communication has been on improving the efficiency in producing content. You'll see conference speakers use the Ford Model T motor car as a metaphor for how technical communicators should take a more engineering approach to technical communication.

Unfortunately, there's been less focus on the improving the value of the outputs we create. Here, we can also earn lessons from the Model T. By 1926, Ford had been overtaken by General Motor as the leading car manufacturer in the USA. Ford had ended up with an efficient way of creating a car that fewer and fewer people wanted to buy. There's a danger we're making the same mistakes as Ford did with the Model T.

While there have been huge leaps in the technology used to create and publish user documentation, it's been quite a while since there were any significant changes to the writing style in technical communication.

Assumptions made by those principles

In most situations, technical communicators adopt a formal tone of voice that is, in most cases, unemotional. The message is mostly authoritative ("Do what I say and you'll be OK"). For many years, this approach to writing technical content has served our audiences very well. Technology has been traditionally expensive, a little bit intimidating, with people afraid of breaking it, and the dominant, succinct approach to writing has provided authoritative, reassuring information.

This approach to writing has many strengths: the information is typically easy to understand, unambiguous, translatable and accurate. It was developed (and is still good) for situations where safety and risk reduction are important, and when the audience is anxious.

Reasons for the rule breaking

We have seen many changes in the last 25 years in the areas of technology, users' relationship with technology, and how technology has sold. These changes challenge some of the assumptions that underlie technical communication.

Changes in technology

Kevin Kelly in his TED presentation, Technology's epic story (Kelly 2009), argues that technology is following an evolutionary path over time, similar to humanity's biological evolution. We can use the five key forces in evolution to create a simple radar chart.

We can describe the shape of technology in the second half of the 20th century as looking roughly like this:

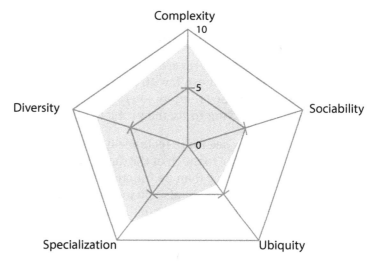

Products tended to specialize in doing one function, and work under different operating systems.

Today, a great deal of consumer technology is now ubiquitous, even mundane. The shape of the diagram for these products would look roughly like this:

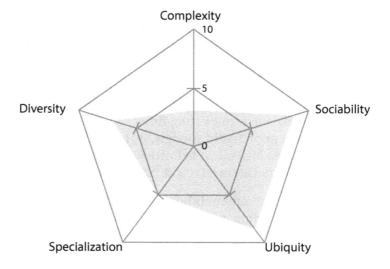

Changes in users' relationship with technology

As technology has become part of everyone's daily lives (particularly web and mobile applications), people's relationship with a great amount of technology has changed.

Many people now regard technology as something that should simply work. When it doesn't, with many products being so much cheaper than ever before, they're more likely to stop using it, even throw it away, instead of trying to fix the problem.

We also now see more technically capable users, because we use technology for so much of the day. There's a greater number of users who want to do more than be functionally competent; they want to master a product. They might also be inclined to tinker and make it do things the manufacturer never intended.

Changes in marketing

Companies spend a lot of money creating glossy brochures and finding the appropriate emotional words to persuade prospective customers to buy their products or services. In the past, this changed once you became a customer. User guides were printed in black and white instead of colour, on cheap paper stock rather than glossy.

However, with web-based services this has changed. Many organizations work on a 'try before you buy' model, and this, alongside the unified experience a web-based product provides, leads them to adopting a single, consistent brand identity at every point of the user experience.

Research shows a large number of people tend to search for the solution to their problem before they buy a product or service. It means we're seeing a new marketing funnel emerge. Google calls this "The Zero Moment of Truth" (Google, 2011), and this change was also reported by Corinna Melville in her article 'A business case for technical communication – facts & figures' (Melville, 2014).

Technical content is becoming the start of the customer journey, because it's providing information that's important to the prospect. This means the content that technical communicators produce must serve an additional function – provide the answers to people's problems in a way that also promotes products and services.

Table 1: Technical communication's role in the customer journey of today

Stage	Content Type
Learn	Solutions to their problems
Try	Product walk-throughs
	Help content
Buy	Help content
Advocate	Best practice
	Tips

Towards a new model of technical communication

Where products no longer fit the 'big, scary and expensive' description, and where users have become more competent, a new model of technical communication may be more appropriate.

Design-led documentation

Some organizations are taking a more design-led approach to user documentation. Citrix has been one of the first documentation teams to adopt this approach.

In 2013, Citrix's Senior Information Experience Manager, Mathew Varghese, described how its technical documentation team began its move towards becoming an 'Information Experience' department (Varghese, 2013). For Citrix, Information Experience is a way of empathising with users to understand their needs and assisting them through the whole customer journey.

Their goal is to provide the right content to the right user, at a time of the user's choosing. This means a more immersive user experience, moving away from confining user assistance to a traditional user manual or help file. This means looking at how content can be delivered in the user interface, in forums, through videos and blogs.

According to technical writing consultant Mark Baker (Baker, 2013):

> The medium is not the sticking point, it is the instruction, the unpleasant exercise of expending mental energy required to learn something of slight or passing interest, and especially of

learning it not from a person, or from experience but from some impersonal and abstract lump of content. They just don't like it.

One voice, many tones

Other organizations are changing the tone of voice and adopting a more conversational approach in certain places in their user guides. For example, the writers at Mailchimp.com first identify the likely emotional state of the reader of a page, and then use a writing tone best suited to that state.

MailChimp is an email newsletter service that enables businesses to send newsletters or email marketing campaigns to a mailing list. They recognize the user's 'state of mind' and deliver content that is best suited to that state. In their style guide for writers (Mailchimp, 2017), it states:

> Before you write content for MailChimp, it's important to think about our readers. Though our voice doesn't change, our tone adapts to our users' feelings.

This means, the tone changes to match the context. Psychologists talk about moving from a dominant to a submissive communication style, but you may find it easier to see it as becoming 'the guide on the side'. In many cases, this means adopting a more informal and submissive tone. By this we mean, you should let them feel the user is in control, and they can choose to do a particular action or not.

A nuanced approach

Does this mean you need to abandon DITA or minimalism? Not necessarily; the new approaches mainly lead to a change in the way introductory and conceptual sections are written. Task information, the procedural steps, usually remain unchanged.

The traditional approach is likely to still be the best approach to take if your users are anxious or frustrated at the point when they start to read your content, or if you are documenting products where mistakes can be expensive or dangerous.

Of course, no idea is ever totally original. Seventeen years ago, Professor Saul Carliner (Carliner, 2000. pp.561–76.) wrote about how to include an affective element in user guides:

Design is an essential ingredient to the success of all these efforts. For example, to develop an online interaction, a technical communicator must not only write the message presented to users, but must first predict users' goals, moods, and motivations, and gear the message accordingly.

If several different types of users encounter the same content, then the communicator must also discover this difference and display a message that's tailored not only to the context and mood, but to the type of user.

Creating a positive relationship with the user

Today, customers have a voice via social media sites such as YouTube and Twitter. What we've seen with the explosion of social media is that people want to be listened to and to share their experiences. They love having conversations: a two-way dialogue.

According to Anne Gentle, author of *Conversation and Community: The Social Web for Documentation* (Gentle, 2012),

> The technical communication world is on the brink of a major cultural shift from one-sided 'documentation'.

By giving users useful, usable and pleasurable experiences, this shift promises to result in more positive relationships with our customers.

Implications for technical communicators

These new approaches can create some challenges for technical communicators, particularly in an international context.

This means clear guidelines need to be given to writers on how to write these more informal topics in a culturally acceptable way.

Plan from the start

So, in order to get it right, whoever creates the content needs to do more than just writing and designing the page layout. It works when it's planned from the start – when the product or service is being designed – and when it is fully tested by users prior to the official release.

It's more about the context

The goal is to fully understand the users, their thoughts and feelings, and to write in a way most appropriate to that context. This means we should be carrying out usability testing wherever possible.

Usability research shows us that an anxious user is less able to think creatively when they encounter problems. In this situation, all the relevant information needs to be close at hand and not overwhelming for the user. We want to create a feeling that there's a 'guide on the side' ready to help them, when they need it. It needs to get people's attention in the right way: attracting and persuading users to approach, persuading them to take particular actions.

This means that when, where, how it appears is as important as the words themselves.

A conversational UI

We'll be seeing more developments with conversational user interfaces, on-boarding screens and micro-content. This is a consequence of users being reluctant to admit they're stuck and going to the help. With Siri, Google Voice and Cortana currently closed off to most developers, we may see conversational user interfaces developed as alternatives to these applications.

Truth Labs's Stelios Constantinides (Constantinides, 2016) has written an article on his experiments with conversational user interfaces.

> Conversational user interfaces (CUIs) are a spoken or written way of interacting with a device. CUIs aren't completely new, but they're becoming smarter, more natural, and – therefore – more useful. Here's where CUIs come in: since users already spend so much time in apps like Slack, Facebook Messenger, and even plain-old email, why not integrate your app inside these platforms?

Constantinides looks at the design process of creating something that doesn't come across as a robot, and isn't as annoying as Microsoft Clippy.

Professor Stephen Pinker (Pinker, 2005) has argued that when we use language to communicate, we are indirectly communicating a relationship between two parties. For example, a technical writer could write as if they were an adult speaking to another adult, or as a parent speaking to a child.

This links in with Ann Rockley's concept of 'intelligent content' (Rockley, 2008):

> Content that's structurally rich and semantically categorized and therefore automatically discoverable, reusable, reconfigurable, and adaptable.

It's still unclear whether this will lead to content being seen as code, or stored in a semantically rich format and inside a content management system.

For conversational user interfaces to work well, they need to be automatically discoverable, adaptable and semantically categorized. Microsoft Clippy wasn't, which may explain why it failed in its purpose.

It's important to recognize that the conversational language is different from written language. We speak differently from the way we write, and this is reflected in how we use messaging apps and authoring tools. Conversations are typically a one-to-one form of communication.

One school of thought is that users will move away from searching using sentences, and, instead, learn to type commands (they will write command-line instructions). In other words, they will begin to think and type more like programmers. I wonder if this might be a bit optimistic.

Implications for translators

Unless it is managed carefully, a more conversational tone is likely to lead to a greater use of idioms in the content. For example, on the "Getting started with Twitter" help page (Twitter.com, 2013), until recently you'd find a section called "GET FANCY: Explore advanced features". "Get Fancy" is a phrase that, I suspect, would tear many translators' hair out.

Users in some cultures might take offence at the content being written in the second person, informal voice (for example, using the "du" form of "you" instead of the more formal "Sie" in German). Even in English, the levels of formality accepted by users in different countries can differ. Contractions or words and phrases, such as 'you'll', can also be problematic to translators.

Obviously, this needs to be planned carefully if you need to localize the text. In the case of Microsoft, when they adopted a more conversational tone in the Office products, they allowed the translators to use an idiom or phrase that was most natural to the target language.

Figure 1: Example of the new empathetic tone in Microsoft Office help

Conclusion

Technical writing, and the nature of the content we provide, is changing, as users' relationship with technology evolves. This means what we teach as best practice in technical communication needs to reflect these changes.

A different tone provides us, in certain situations, with an opportunity to use the power of positive emotional engagement to create more passionate and loyal users as well as better users of our products. It's not just about the words or the page layout, however. This approach needs to be carefully designed to the users' situation, with great consideration to the context in which it appears. In many situations it is not suitable. User assistance still needs to be clear and unambiguous, and safety and risk consideration may preclude this approach.

With web analytics, user feedback and usability testing, we have the ability to analyze user behaviour and gain a more detailed understanding of users. We have the opportunity to identify the situations where the nature of the content we provide – the tone and how content is delivered – should change so that it better meets the users' needs.

References

Baker, M. (2013). Write for people who actually read documentation. [online]. Available at: http://everypageispageone.com/2013/02/25/write-for-people-who-actually-read-documentation/ [Accessed 6 Dec 2016].

Carliner, S. (2000). Physical, cognitive, and affective: A three-part framework for information design. *Technical Communication* (47.4). pp. 561–76.

Constantinides, S. (2016). Where does conversational UI leave design? [online]. Available at: https://medium.com/truth-labs/where-does-conversational-ui-leave-design-7044c395be9f#.eez9gl4w [Accessed 6 Dec 2016].

Gentle, A. (2012). *Conversation and Community: The Social Web for Documentation*. XML Press. 1st Edition. Laguna Hills, CA. XML Press. Back cover page.

Google.com (2011). Zero Moment of Truth (ZMOT). [online] Available at: www.thinkwithgoogle.com/collections/ zero-moment-truth.html [Accessed 6 Dec 2016].

Huxhold, M. and Luther, A. (2013). Produkt und Lernvideos als ideale Erg nzung zur klassischen Dokumentation. [online] Available at: http://conferences.tekom.de/fileadmin/tx_doccon/slides/364_Produkt_und_Ler nvideos_als_ideale_Erg_nzung_zur_klassischen_Dokumentation.pdf [Accessed 6 Dec 2016].

Kelly, K. (2009). Technology's Epic Story. [online] Available at: www.ted.com/talks/kevin_kelly_tells_ technology_s_epic_story [Accessed 6 Dec 2016].

MailChimp.com. (2017). Writing Goals and Principles. [online]. Available at: http://styleguide.mailchimp.com/writing-principles/ [Accessed 20 Feb 2017].

Melville, C. (2014). A business case for technical communication – facts & figures. [online] Available at: www.tcworld.info/e-magazine/ technical-communication/article/ a-business-case-for-technical-communication- facts-figures/

Pinker, S. (2005). What our language habits reveal. [online]. Available at: https://www.ted.com/talks/steven_pinker_on_language_and_thought? [Accessed 6 Dec 2016].

Rockley, A. (2008). What is intelligent content? [online]. Available at: http://www.rockley.com/articles/What%20is%20Intelligent%20Content.pdf

Twitter.com. (2013). Getting started with Twitter. [online]. Retrieved from https://support.twitter.com/groups/50-welcome-to-twitter/. Available at: https://debrasanborn.files.wordpress.com/2013/10/twitter-101-info.pdf [Accessed 6 Dec 2016].

Varghese, M. (2013). Applying Design Thinking to the traditional information development process to deliver great user experience. [online] CIDM Best Practices. Available at: http://dev.infomanagementcenter.com/publications/best-practices-newsletter/2013-best-practices-newsletter/information-experience-design/ [Accessed 6 Dec 2016].

Wikipedia.org: The Free Encyclopedia. (2016). Minimalism (technical communication). [online] Available at: http://en.wikipedia.org/wiki/Minimalism_(technical_communication) [Accessed 6 Dec 2016].

5

"No manuals" – writing user interface copy

Andy Healey

Abstract

As software becomes easier to use, some technical writers are embracing the opportunity to provide 'front-line' help by designing and writing user interface (UI) copy.

Drawing on their inherent language expertise and empathy with users, technical writers are embedded into product design teams, helping to create simple and usable software.

They're using skills and methodologies usually associated with other job functions including:

- Corporate branding, and tone and voice
- User research and A/B testing
- Nudge techniques
- UI design standards

From button labels to tooltips, this chapter will introduce you to producing 'invisible' help to guide users through their journey.

Keywords   copy; user experience (UX); voice; usability; design

Introduction

In a world of apps and mobile, business software companies have spent the last few years promoting concepts such as *consumerization* and *gamification*, and have been making their products easier to use. At the same time the idea has arisen that "in ideal interfaces help is not needed because the interface is learnable and usable" (Porter, 2012). Or as you might have heard someone senior in your organization say, "No manuals!".

It's true that some software is so intuitive and simple that not only are no manuals needed, but words are not needed either – think Angry Birds or, further back, Pong. And with advanced UX techniques and universalized palettes, users can often guess correctly which button they need to click. At the time of writing, it's usually the green one. Unless it's the blue one.

Traditionally, the user interface (UI) copy was consigned to the end of the development process. With no content strategy, a technical writer or product designer would be expected to put the final touches to a product by 'adding a few words'.

Yet these apparent negatives of 'no manuals' and lack of priority attached to UI copy actually conflate into a positive opportunity. With business software, processes and actions can't be trusted to a best guess, and the words used in an interface are the glue that holds the elements of a product together and gives them meaning. To have no (or minimal) manuals means that the product must be usable. To be usable the UI copy can't just be bolted on at the end, but needs to be an integral part of the design process.

Writing UI copy is currently undergoing a renaissance, with companies such as Google actively recruiting for UX writers.

They understand that people use business software not because it's pretty but to get a job done. If people can't use your software quickly, a significant number won't search the help, they'll find a simpler alternative. So you need simple, clear words that tell them how to use it.

A word is worth 1000 pictures

Apple is synonymous with the concepts of placing design and user experience (UX) at the forefront of product design. Yet as far back as 1985 they recognized the importance of UI copy, and the Apple human interface group had the motto "A word is worth a thousand pictures" (Tognazzini, 2000).

The human brain has had several thousand years of semantic association to support the use of language. We have long-established reference patterns with words that haven't been established to the same extent with UI images.

Sometimes an image alone will make sense to us independently, for example, thanks to Facebook's ubiquity, the thumbs up on a Like button is a symbol with a clearly understood action for most of us.

Figure 1: Like buttons

But it might not be understood by everyone: for example, someone who isn't a Facebook user, or the 36 people out of 1200 surveyed from 40 different countries for whom the thumbs up represents a sexual insult (Morris, Collett, Marsh, and O'Shaughnessy, 1979).

Understanding the value of text, Facebook usually clarifies the Like icon with some copy that says "Like". There are endless other interactions we can have with software that don't have such an easily relatable icon. So even with simple or common UI controls, "text labels are necessary to communicate the meaning and reduce ambiguity" (Harley, 2014). Icons without supporting words are usually meaningless, as in the example in Figure 2.

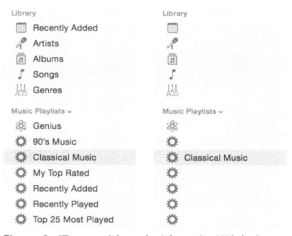

Figure 2: iTunes with and without its UI labels

Good UI copy not only helps people to complete a single task or action as easily as possible. It builds trust that future tasks will also be completed with the same ease. From our personal relationships, to which politicians we vote for, trust is a key factor in our decisions. By using the right words you can engage users to trust your software, and consequently continue to use it rather than turning to your competitor.

Finding your writing style

Your writing style (or tone and voice) is about how your company speaks to its customers and the language that it employs. In many cases this will be directly aligned to a company's broader branding guidelines. The brand represents the people behind a product, who they are and what they believe, and can be used to distinguish a company from its competitors.

MailChimp is a well-known example of a company with a clearly identifiable brand and writing style, further enhanced by making it public-facing at http://voiceandtone.com/. As soon as you start using MailChimp software, the UI copy, coupled with the UX design, presents you with a distinct personality.

Welcome to MailChimp

Alright, let's set this up! Tell us a bit about yourself.

First name

Last name

Figure 3: MailChimp's sign up screen

As previously mentioned, user trust is essential, and one method to achieve it is to have a coherent and recognizable identity. This is something Cisco were intent on doing when they ran a project to define their brand language. "We heard from customers, partners, even our own employees – people didn't understand us. Our new brand voice changes all of that." (*Engage Customer* magazine, 2014)

Tone and voice is a big thing right now, with plenty of consultancies willing to help you find the right voice for your organization. If your company has branding guidelines, then you have a clear starting point for defining your product tone and voice.

However, unless you're looking to do something particularly sophisticated or different (as with MailChimp), then by following a few simple guidelines you can develop a writing style to use in your UI copy.

Most of the time, finding a style equates to keeping things simple, clear, and non-offensive. Are you looking to create a unique identity with humour, or to help your users through clear language? Neither one is 'correct', but in most cases it will be the latter.

Figure 4: Handbrake – fun and memorable

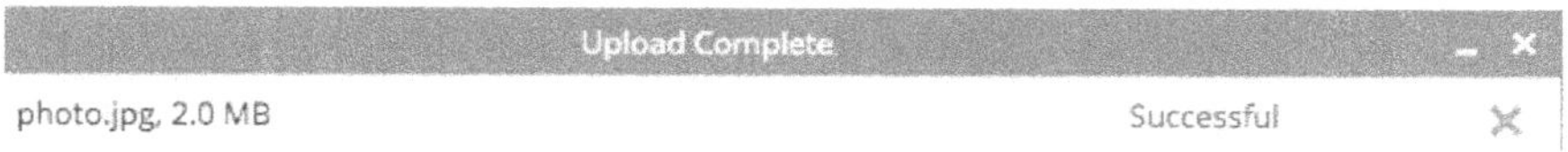

Figure 5: Alfresco demo app – clear and simple

Fun UI copy can be refreshing, but if you're producing business software then remember that 80% of people prefer sentences written in clear English (Morris, 2014).

There's plenty of research: for example, Microsoft's presentations on 'No more robot speak' (Kim, 2014), showing that a clear majority of people (including developers and other 'techies') want the product voice to be clear, non-technical, and not to look like it's trying too hard. As Charlie Brooker complains "No one uses terms like 'sync' in real life. Not even C3PO. If I sync my DVD collection with yours, will I end up with one, two, or no copies of Santa Claus the Movie?" (Brooker, 2011). Put simply, the words shouldn't be a distraction from the user's task in hand. The advertising and copywriting industries learnt years ago to talk to people in their own language, and now the software industry is catching up.

Select an item to read.

Click here to always select the first item in the list

Figure 6: Microsoft Outlook, no more robot speak

So when you write for your product there are three overridingly important factors:

1 Consistency – Create a pattern of usage users can learn from

2 Empathy – Understand your users situation and help them

3 Simplicity – "Simplicity is the ultimate sophistication"

Simple is not the same as simplistic, and audiences do vary. A learning app for children will have different language to a learning app for lawyers, but lawyers still want it simple. Because it was costing them business, IBM recently simplified their cloud legal documents from 30 pages to 2 (Cohen, 2015).

And think about how your audience is geographically distributed. If you have a feature that can be on or off, then 'turn off' makes more sense to an American user base than 'switch off' does. And either one of these has less potential to provoke an emotional response than 'disable'.

Aside from user preferences, many governments now actively mandate that any writing done is in Plain English. In 2010 the US Government mandated that all federal communications (including websites) must adhere to the Plain Language bill: "The best way to do this [provide effective communications] is by using common words and working with natural reading behaviour" (FDA, 2011). The UK government also state that "Plain English is mandatory for all of GOV.UK" (GOV.UK, 2014).

Whoever your audience is, and whatever style you use, your writing shouldn't be a challenge, it should be cognitively invisible.

Design standards for UI writing

When you've decided on your writing style, the next step is to formalize a set of design standards for UI writing; effectively a writing style guide. Google, Apple, and Microsoft all have good examples of these.

Having a set of formal writing guidelines means that you have clear boundaries against which you can benchmark existing copy and create new copy, and lets you clearly define what is (and isn't) correct use of language.

Your UX team will likely have a set of design standards to ensure that you have consistent branding throughout your products. This may include high-level stylings such as flat design or floating palettes, as well as design specifics such as colour palettes, fonts sizes, or axes to be used for 3D-space. Your writing guidelines should stand alongside these, or even better, be incorporated directly into them. This is a great opportunity to be creative and set in stone a reference that can be used to standardize how your products use language.

Below are a few examples taken from Alfresco software's writing guidelines (Alfresco, 2016) showing the kind of things you can do:

1 Show how to empathize with users (Figure 7 on page 96 and Figure 8 on page 96).
2 Define which words to use (Figure 9 on page 97).
3 Create linguistic patterns to be used in your error messages (Figure 10 on page 97).
4 Instruct in the use of active writing and nudge techniques (Thaler and Sunstein, 2008) to encourage users to move on to the next step (Figure 11 on page 97).
5 Demonstrate the usage of plain language (Figure 12 on page 98).

Many companies have made their writing guidelines freely available, and much of it is tried and tested, so there's no need for you to design something from scratch. Look through what's already been produced and adopt what works for your company.

Compile all this into an official 'document' in whatever format works best for you. Whether it's PDF on your company intranet or a series of web pages, the key thing is that this must be visible, and known about. And of course they should be flexible and versionable, as UX fashions can and do change quickly.

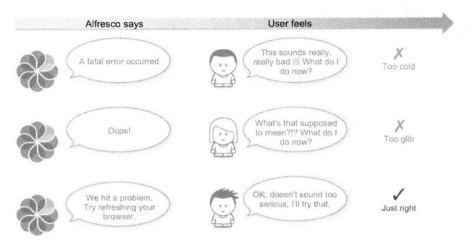

Figure 7: Alfresco Software's writing guidelines: how to empathize with users

Tone - 4 tips

Who?
Think about the person you're writing for, the Alfresco product they're using, and the level of knowledge they have. A developer can handle more technical language than an end-user, but they both want it to be kept as simple and clear as possible. Empower users by making them feel knowledgeable, rather than assuming a knowledge-level they may not have.
Who this is for? How do they feel? What are they trying to do? What do they need to know?

How?
Write it how you'd say it. When you write read it out loud. Is it clear? Does it make sense? Does a real person talk like this? Think of UI language as a two-way conversation between the user "you" and Alfresco "we".

Be consistent
If a term has already been established, then use it, don't try to reinvent it. Stick to the same term within a process, for example, don't talk about both "deleting" and "removing" if you mean the same thing. If you're writing about something new then be creative, but stick to plain language.

Be professional
Alfresco products are enterprise products and the language needs to reflect this. Strike a middle ground between over-formality and over-friendly. Write how you'd speak to a customer.

Figure 8: Alfresco Software's writing guidelines: tips on tone

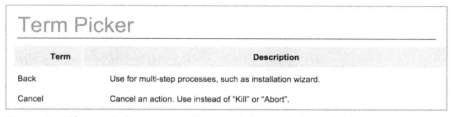

Figure 9: Alfresco Software's writing guidelines: define which words to use

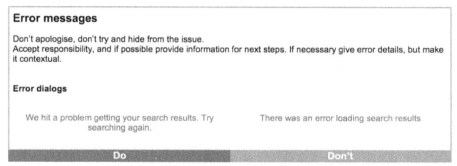

Figure 10: Alfresco Software's writing guidelines: create linguistic patterns to be used in your error messages

Figure 11: Alfresco Software's writing guidelines: instruct in the use of active writing and nudge techniques

Write simply and directly

Use plain language. Remove unnecessary words when possible, as they slow down the reading process and open up opportunities for confusion.

Upload new version   Would you like to upload your new version?

Do	Don't

Figure 12: Alfresco Software's writing guidelines: demonstrate the use of

plain language

Once they're done, you're only just getting started. You may already have established some supporters at a senior level for the project to get this far. You now need to evangelize it throughout your entire organization. If this is to be used for benchmarking then it needs to be accepted and used. You'll need to invoke a change management process so that the writing guidelines become part of your development standards and quality control process.

Integrating writing into the design process

It's commonplace for UI copy reviews (if there are any) to be at the final stage of development, after the coding has already been done. The UI copy may well have been written by the software engineers. When up against release deadlines, bad writing and typos are rarely seen as a 'blocker', so they don't get fixed.

This approach carries with it a range of hidden costs that don't tend to show up on financial charts, and so are often ignored. There are, however, implications on multiple cost centres in an organization. If the copy in a product isn't clear then customers will call your support team for help. This costs time for the team. Support will raise a bug fix request with engineering. If they fix the bug then this takes time away from something else that they could have been doing. And of course this is the second time the engineer has gone into the code for this piece of copy. Once that's done then the copy needs translating again.

Figure 13: Cost cycle of no review

Instead of each of these largely unknown and difficult to calculate cost blockers, a more clearly defined process can be used to calculate costs up front, and reduce the impact on additional teams.

In an ideal world, the copy would be created completely in parallel with the initial designs. The words not an afterthought, but an integral part of a solution that solves a user problem. In reality, UX designers often aren't that into the writing part of design, so concentrate on the visual side. This means that copy that was effectively thrown in as a placeholder ends up in the final product – see Figure 13. Either that or your designers will use an actual placeholder, such *Lorem ipsum*.

This Latin-based placeholder text has been used in publishing since the 1500s, and in software since PageMaker first used it in the mid-1980s. Surprisingly, many designers still use *Lorem ipsum* as a placeholder, despite the numerous articles and evidence showing why it's a bad process to follow:

- It can cause confusion when previewing designs
- It cognitively distracts from the design
- It can break the design (see Figure 14) – it's unlikely that the actual content will fit the same space as a placeholder
- It might accidentally end up being more than a placeholder

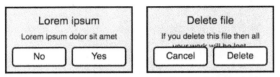

Figure 14: Lorem ipsum breaking the design

A simple way to counter these potential pitfalls is to work with your UX team as they design wireframes, or at least to have a process for reviewing their wireframes *before* any development takes place.

In this 5-step example:

1 The product manager creates a story for the design.
2 The UX designer produces a design prototype or wireframe.
3 The user assistance team reviews the design with UX and agrees on UI copy.
4 The design is validated by product management and customers/users (and can be moved back to the previous stages if this results in change requests).
5 The design is ready for development.

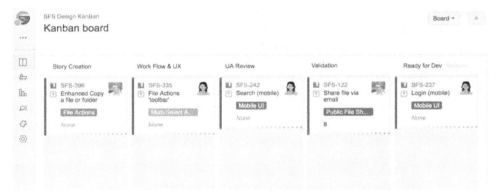

Figure 15: Sample Jira design flow with User Assistance (UA) review

This simple process flow works whether you're using Jira or not, and you can easily tailor it to fit your design and development cycle.

Of course, the way the product works might change from the original designs due to engineering limitations, customer feedback, or other reasons, so you can't create definitive final text. What you should be aiming to do is create 'proto-content'. This content should be as close as you can get to what the final copy will be.

To create this content you need to interview and work with the designers and product managers. Determine exactly what the product or feature will be doing. Check competitor products to see what they use, or if there's an industry-standard term that should be used. If you have a new feature then avoid making up words or trying to be too clever – I once had to ask an engineer what the name of his new feature meant, and he smiled "How much do you know about polymer chemistry?" The feature and product had absolutely nothing to do with polymer chemistry.

Using proto-content means that your designers have a better feel for the end product. Rather than seeing text as just a component that can be bolted on, it's a tangible part of their wireframes, adding value to demonstrations, and avoiding costly changes later.

Defining and using a UI copy taxonomy

You've decided with your team what copy is needed. You have writing guidelines to keep you on message. Your next question is *how* to present the copy. Good help is invisible. If it's truly integrated into the product then users see an easy-to-use product, product management sees a successful user experience, and senior management sees 'no manuals'.

The primary purpose of traditional online help and manuals is to help users understand and complete a task. It's quicker for a user to see how to complete a task for themselves, rather than searching for how to do it in another location such as a help system or Google. This can mean they need concept, task, and reference information, all in one screen.

As far back as 1995 Jakob Nielsen advocated that "it is better if the system can be used without documentation" (Nielsen, 1995). And with Windows 95, Microsoft attempted to do this when they introduced hover tooltips for the toolbar.

Figure 16: Windows 95 tooltips

Every piece of UI copy you create should have this same purpose – to keep the user in context, in your software, by helping them to complete their task.

And while each piece of UI copy serves towards this same goal, there is an array of formats we can use to give information. This lets us vary the level of information we provide, depending on the complexity of the task. Figure 17 shows multiple ways that UI copy can be presented.

There are:

- Button labels
- Section titles
- Mouse–over tooltips
- Clickable tooltips
- Empty state placeholders
- Instructional task descriptions

Even the two on-screen search boxes tell you exactly what you can search for in each box.

These different types of 'microcopy' work together, from full task descriptions to button labels, so you can be adaptive based on the level of information that's needed to complete a task.

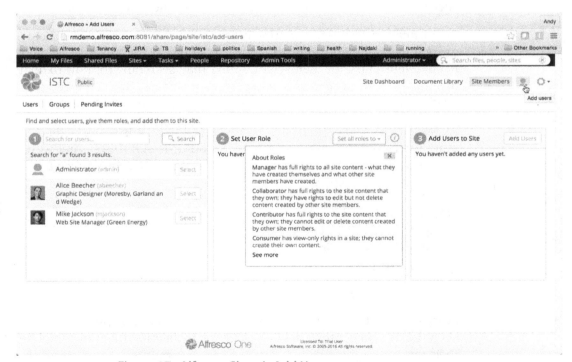

Figure 17: Alfresco Share's Add Users screen

A copy taxonomy defines which copy types to use and when. The options available to you in designing a taxonomy will be affected by how your copy is implemented in the product (for more on this, see 'Implementing and managing UI copy' on page 104).

For each piece of copy you should consider where the user is on their journey and how much information they need. Users only read 20% of the words on a website (Nielsen, 2008), so don't overburden them, and make each word count. Use progressive disclosure to gradually reveal information, as the user needs it: "By disclosing information progressively, interaction designers reveal only the essentials and help users manage the complexity of feature-rich sites or applications." (Spillers, 2004).

Each new piece of copy should be matched to the appropriate type from your copy taxonomy. In most cases this will be straightforward, but there will be times when you need to decide if, for example, a line of help copy is best placed on screen below a field, or would be a distraction and is best hidden behind a 'More Info' button. Or how to repurpose a mouse-over tooltip for mobile. Decide if the user needs that information in front of them for their next action.

Figure 18: Does the user need all of this information at this point?

By breaking down each section of UI into granular chunks such as buttons, fields, section titles, and so on, the boundaries of your copy taxonomy are largely shaped for you. Following these boundaries creates consistency and familiarity for ease of use.

Table 1: Sample copy taxonomy for a field UI component

Type	Information level	Copy length
Field label	Essential for task completion	1 or 2 words
Field description	Additional info to field label. Required if label isn't universally clear.	1 short sentence (if required)

Table 1: Sample copy taxonomy for a field UI component

Type	Information level	Copy length
Field info button copy	Optional for users who want to make a more informed decision. Rarely required.	1 sentence to several paragraphs, with possible link to online help.
Field error message	Essential for task completion	1 or 2 sentences. Technical detail less important than detail on how to resolve error.

During the design review you should have already established what UI copy is needed, and you have clear writing guidelines to follow as to how content should be written, such as which words are used on buttons in given situations. You (and/or your product designers) need to anticipate what the user will need to know at each stage of your product, and provide that information in the most context-appropriate format.

Implementing and managing UI copy

How you implement your UI copy – actually have it appear in the product – is usually a product management decision, and may be restricted by your software environment.

UI copy strings might be hard-coded into a product, making them difficult to access, modify, or localize. This method also takes little consideration of accessibility.

More common is for UI copy to be stored in sets of property files or similar, alongside all the other files that when compiled create your product. This means that you can access them, and make changes using your preferred text editor. Different types of properties files or software build processes do have different requirements, for example, whether you use straight, curly, or double apostrophes, so be clear on any potential issues before you start any editing. Over 90% of software companies use some form of version management software such as Perforce or Github. With these tools you can pull, push, and merge changes, just as your developers do with other areas of the code. This may take a little getting used to, but it is an effective method. Any copy changes can be 'outsourced' from developers to the writers. This not only reduces the time and process to get changes made, effectively cutting out the developer middleman, but also means that the writer can be sure the copy is as they intended. And it wins you friends in development by taking an unwanted task on for them.

```
26  ## Dialog labels
27  label.new-rma_recordCategory.title=New Record Category
28  label.new-rma_recordCategory.header=Record Category Details
29  label.new-rma_recordFolder.title=New Record Folder
30  label.new-rma_recordFolder.header=Record Folder Details
31  label.new-rma_hold.title=New Hold
32  label.new-rma_hold.header=New Hold Details
33  label.new-rma_unfiledRecordFolder.title=New Folder
34  label.new-rma_unfiledRecordFolder.header=Folder Details
35  label.new-rma_nonElectronicDocument.title=Non-electronic Record
36  label.new-rma_nonElectronicDocument.header=Record Details
37
38  ## Drop-down Menus
39  menu.selected-items=Selected Items...
40  menu.selected-items.deselect-all=Deselect All
41
42  menu.selected-items.accession=Accession
43  menu.selected-items.accession-complete=Accession Confirmation
44  menu.selected-items.copy-to=Copy to...
45  menu.selected-items.cutoff=Cut Off
46  menu.selected-items.delete=Delete
47  menu.selected-items.destroy=Destroy
48  menu.selected-items.export=Export
49  menu.selected-items.link-to=Link to...
50  menu.selected-items.file-to=File to...
51  menu.selected-items.move-to=Move to...
52  menu.selected-items.reviewed=Reviewed
53  menu.selected-items.transfer=Transfer
54  menu.selected-items.transfer-complete=Transfer Confirmation
55  menu.selected-items.undo-cutoff=Undo Cut Off
56
57  ## Pop-up Messages
58  message.file.type=Are you filing an Electronic or Non-electronic Record?
59  message.file.type.title=Filing Type
60  message.multiple-delete.failure={0} items successfully deleted. {1} items couldn't be deleted.
61  message.multiple-delete.success=Successfully deleted {0} item(s)
62  title.multiple-delete.confirm=Multiple Delete
63  message.multiple-delete.confirm=The following {0} items will be removed from the file plan.
64  message.multiple-delete.please-wait=Deleting records...
65  message.new-unknown.error=We couldn't find type ''{0}''.
```

Figure 19: Properties files in Github

There are also an increasing number of solutions that you can 'bolt-on' to your product to give additional options for creating tooltips, product-walkthroughs, polls, and other options for inline help that may not be available in your development environment. Current offerings include WalkMe, Appcues, and Pendo.

They don't replace the need for the basics such as button labels that are integral to your product, but they do offer an extra array of taxonomy options without the need for any development. For example, some of these products have integrated analytics. They can see when a specific user hasn't used a certain feature in their first, say 20 uses of your product, and you can write copy for a popup that explains why they might want to use it and what benefits it will give them. This is a classic way of using nudge techniques to subtly help users get the most from a product.

The implementation of the copy is ultimately an area where the writer is largely subject to the development environment. However, your informed recommendations – "We shouldn't hardcode the copy", "I can edit these files myself", "Maybe we can add feature reminders without the developers getting involved" – do add value to the development lifecycle and subsequent user experience.

Validation

As UX has splintered into various sub-categories, one that has gained increasing value is UX research and validation. Designing a pretty (and well-written) product is great, but is it what people want to use and does it address the issues they're trying to solve?

Writing isn't a binary art, and there'll be different opinions on what the best wording for a piece of copy is, even when following company guidelines. So why not put these options in front of users and customers and let them say what works?

If your company already has a UX Research team then it's time to add terminology to the list of research tasks. If you don't have existing research facilities available, then it's fairly straightforward to set up some basic validation yourself.

The two most important research tools to use are concept validation and A/B testing.

Concept validation

Concept validation involves providing sketches, wireframes, or even interactive prototypes. As discussed already, you should have the writing process fully integrated into the design stage. Ideally, the designs will be shown to customers or partners, and this will usually be done by a product manager or product evangelist. If your company isn't conducting validation exercises prior to development, then now's the time to start!

You'll find out what works and what doesn't, and the design can be improved and iterated on using this feedback. This is the ideal time for finding out whether your copy has the clarity for users to complete tasks, and just as importantly, whether it is audience-appropriate. For example, do the terms you use make sense in a particular industry.

As concept validation is focused on the overall design rather than just the words, you may want to get more specific, and this is where the A/B testing comes in.

A/B testing

A/B testing (or split testing) is where you give participants a choice of two similar options, and see which one they prefer, or which one is more successful in task completion. This is a popular marketing and business analysis tool, particularly for websites to optimize conversion rates, and it can be as simple as just changing a few words. Veeam Software is a great example of how changing one word on their website resulted in 162% increase in click-throughs (Deswal, 2012).

A/B testing is highly effective when you're looking at specific areas such as a choice between copy options. According to Jakob Neilson "there's close to 100% probability that you'll choose the design variation that makes the most money [or is the most usable]."(Neilson, 2012) It's also very easy to conduct. If you have access to customers or users then that's perfect. Present them with two options and see which one they prefer. There are also plenty of online tools available, such as Validately, that you can use at low cost.

And don't worry about getting large numbers to check this, just five or so participants from the appropriate user groups is all you need (Nielsen, 2000).

An additional option if you have a lot of existing content that needs improvement and you don't know where to start is to consider conducting a user survey. You can do this simply and for free using tools such as SurveyMonkey.

Product Copy

1. How often do you use our product?

○ Daily

○ Weekly

○ Monthly

○ Never

2. How easy it it to use?

Very easy	Easy	Average	Difficult	Very difficult
☆	☆	☆	☆	☆

3. Is there anything you really like?

4. Is there anything you really don't like or have difficulty with?

5. Do you see any terms you don't understand or that are confusing in the product?

○ Yes

○ No

○ Tell us what terms we could improve

6. Could changing the words we use in the product improve your experience?

○ Yes

○ No

Done

Figure 20: Sample copy survey using SurveyMonkey

With a few simple questions, you can find the exact areas to prioritize for improvement, as well as having validation that this is indeed a customer requirement. In a recent similar survey I conducted, almost 50% of users said that changing the UI copy would improve their product experience, and with their feedback we knew exactly what to change.

Whichever validation method or methods you choose to use, you'll very quickly establish compelling evidence and business justification as to which option you should go with, moving from 'we think' to 'we know'.

Compliance and tolerance

With writing guidelines established, you're in a good place to know when copy is up to scratch and when it isn't. Running a compliance audit is a fairly straightforward way to identify how well you're doing and how much room for improvement there is.

To run a compliance audit, just go through the existing copy and see how close it is to your guidelines. You could do this on a product basis, a feature basis, or by evaluating each individual file that holds your UI copy.

For example, you may have a feature that has 1000 lines of UI copy. It's a fairly simple task to go through these 1000 lines and identify how many meet your writing guidelines and how many don't. You can then build in a tolerance level.

If 950 lines in the file adhere to your writing guidelines, and only 50 lines in this file don't (unhelpful error messages, unclear field labels, and so on), then the feature would be 95% compliant. You can then set a tolerance. Some companies may see 95% as sufficiently compliant with their guidelines. Others would see it as not compliant, especially if those 50 lines were causing repeated customer issues.

Wherever you decide to set your tolerance levels, running compliance audits is a crucial way of identifying how much resource is required to achieve compliance. It's also a very clear identification of how you have improved the product and its usability.

Taking a product from non-compliance to compliance can be very important. As mentioned earlier, some countries now mandate that software must conform to Plain Language, and usability and accessibility feature in ISO-9241. Some countries also have legislation stating that some writing types (such as insurance policies) must fall within a certain Flesch–Kincaid range. Flesch–Kincaid is the most widely used readability checker which scores how readable your copy is. Requirements vary depending on the target audience, but a score of 65 (from a range of 1 to 120) is a good aim for business writing.

Flesch–Kincaid is actually built into Microsoft Word. Having spent time and effort defining guidelines, integrating writing into your development process, and validating with your customers, this is the pay-off. You can just copy and paste your content into Word for quick confirmation of its readability.

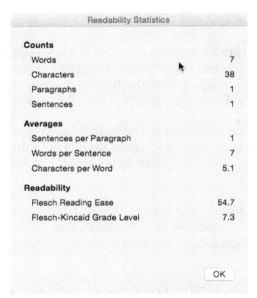

Figure 21: Using Flesch–Kincaid in Microsoft Word

Alternatively, you can use more advanced software, such as Acrolinx, which will automate some of this for you, as well as giving feedback on grammar, tone, and, if you align it with your UI writing standards, ensuring correct terminology usage.

Projections for the future

Apps are going to become simpler and smaller. There'll be less requirement for end-user help, and manuals (end-user ones at least) will in many cases become a thing of the past. Not in all cases; for example, accounting software can only be simplified to a certain level, but the emphasis will be on providing contextual help. Tools such as Swagger are already enabling help to be more contextual even for API documentation.

Writers will no longer translate engineers' notes into readable language and push it out to users. Everything they write will be validated up front to ensure successful products. "Without knowing what your particular target audiences are doing and getting away from stereotypes, you'll be building things that are just a gamble." (Kendall and Weaver, 2015)

The visual appearance of software will become more homogenized, as usage of design pattern libraries such as Google's Material Design becomes more widespread. The emphasis will be less on the design, and more on the information software can provide. The likes of Twitter and

Instagram (which generates 3.5 billion likes per day) are pointing the way to where "notifications are the future app interface" (Acharya, 2015) and there will a "rise of conversational UIs" (Connolly, 2016).

The example set by leading brands such as Google and Facebook for employing UX writers, language designers, and content strategists will spread, and writers who would once have been writing end-user help, will be producing conversational copy for apps and notifications. In the long-term, the increase of artificial intelligence (AI) algorithms will enable at least some of this to be automated.

Conclusion

Writing UI copy isn't about making users happy; its purpose is to stop them from becoming unhappy. It's taking traditional technical writing approaches such as minimalism and empathetic writing, and contextualizing them within the product. As software design becomes less about product design and more about experience design, an increased emphasis on UI copy improves the user experience.

The principle of 'no manuals' serves to bring writers into the design team. Much of what was previously written in help or manuals is actually better placed in-product, so that the user never has to ask a question in the first place.

If you're working on a brand-new product then setting up the design process is relatively straightforward. If your product has thousands of strings of existing copy, it may seem daunting to optimize and align them, but by following a pragmatic, piecemeal approach you can address the recommendations in this chapter one at a time. Start off by approaching the designers on new features and offer to review the copy they've produced. It's highly likely that you'll improve the usability of that feature.

Remember that the designer, product manager, or developer who wrote the copy have been immersed in the feature and know what it does (or what it should do). For exactly the same reasons as technical writers write the help, so they are also ideally placed to write the UI copy. Your perspective is essentially that of a user, and your words will help the feature avoid what Stephen Pinker refers to as the curse of knowledge: "It simply doesn't occur to the writer that her readers don't know what she knows…And so she doesn't bother to explain the jargon, or spell out the logic, or supply the necessary detail." (Pinker, 2014)

References

Acharya, A. (2015). New Resource: Notifications are the next platform. Available at https://techcrunch.com/2015/04/21/notifications-are-the-next-platform/ [Accessed 8 Feb 2017].

Brooker, C. (2011). New Resource: I don't hate Macs, but they do give me a syncing feeling. Available at: https://www.theguardian.com/commentisfree/2011/feb/28/charlie-brooker-pfroblem-with-macs [Accessed 8 Feb 2017].

Cohen, M. (2015). New Resource: Memo to File: Keep it Simple. Available at: http://legalmosaic.com/2015/03/16/lawyers-keep-simple/ [Accessed 8 Feb 2017].

Connolly, E. (2016). New Resource: Design is a Converation. Available at: https://blog.intercom.com/designing-conversational-interfaces/ [Accessed 8 Feb 2017].

Deswal, S. (2012). New Resource: How changing a single word increased click through rate by 161%. Available at: https://vwo.com/blog/increase-click-through-rate/ [Accessed 8 Feb 2017].

Engage Customer magazine (2014). New Resource: How Cisco Changed its Brand Language and the Customer Experience. Available at: https://issuu.com/cenict/docs/ec_mag_jan_14_lr__1_/12 [Accessed 8 Feb 2017].

FDA (2011). New Resource: Federal Plain Language Guidelines. Available at: http://www.fda.gov/AboutFDA/PlainLanguage/ucm346268.htm [Accessed 8 Feb 2017].

GOV.UK (2014). New Resource: Writing for GOV.UK https://www.gov.uk/guidance/content-design/writing-for-gov-uk [Accessed 8 Feb 2017].

Harley, A. (2014). New Resource: Icon Usability. Available at: https://www.nngroup.com/articles/icon-usability/ [Accessed 8 Feb 2017].

Healey, A. (2016). New Resource: Alfresco writing guidelines. Available at: http://www.docs.alfresco.com/writing-for-alfresco [Accessed 8 Feb 2017].

Kendall, A. and Weaver, M. (2015). The Customer Model of IA & Content Strategy. Available at: http://content26.com/blog/the-customer-model-of-ia-content-strategy-interview/ [Accessed 8 Feb 2017].

Kim, D. (2014). No. More. Robot. Speak. – Our journey to a (way) better Microsoft Voice and why it matters In: TCUK 2014. Brighton.

Morris, D, Collett, P, Marsh, P, O'Shaughnessy, M, (1979). Webified by Bernd Wechner: Gestures: Their Origin and Meanings (1st ed). Available at: https://www.scribd.com/document/89553971/Bernd-wechner-info-Hitchhiking-Thumb [Accessed 8 Feb 2017].

Morris, M. (2014). New Resource: Clarity is king – the evidence that reveals the desperate need to re-think the way we write. Available at: https://gds.blog.gov.uk/2014/02/17/guest-post-clarity-is-king-the-evidence-that-reveals-the-desperate-need-to-re-think-the-way-we-write/ [Accessed 8 Feb 2017].

Nielsen, J. (1995). New Resource: 10 Usability Heuristics for User Interface Design. Available at: https://www.nngroup.com/articles/ten-usability-heuristics/ [Accessed 8 Feb 2017].

Nielsen, J. (2000). New Resource: Why You only Need to Test with 5 Users. Available at: https://www.nngroup.com/articles/why-you-only-need-to-test-with-5-users/ [Accessed 8 Feb 2017].

Nielsen, J. (2008). New Resource: How Little Do Users Read? Available at: https://www.nngroup.com/articles/how-little-do-users-read/ [Accessed 8 Feb 2017].

Nielsen, J. (2012). New Resource: A/B Testing, Usability Engineering, Radical Innovation: What Pays best? Available at: https://www.nngroup.com/articles/ab-testing-usability-engineering/ [Accessed 8 Feb 2017].

Pinker, S. (2014). The Thinking Person's Guide to Writing in the 21st Century. London: Penguin, p. 61.

Porter, J. (2012). New Resource: Principles of User Interface Design. Available at: http://bokardo.com/principles-of-user-interface-design/ [Accessed 8 Feb 2017].

Spillers, F. (2004). Progressive Disclosure. [online] In: M. Soegaard and R. Friis Dam, ed., The Glossary of Human Computer Interaction. Available at: https://www.interaction-design.org/literature/book/the-glossary-of-human-computer-interaction/progressive-disclosure [Accessed 8 Feb 2017].

Thaler, R and Sunstein, C. (2008) Nudge: Improving Decisions About Health, Wealth, and Happiness (1st ed). New Haven: Yale University Press.

Tognazzini, B. (2000). New Resource: Trends: The Evolution of the Interface. Available a:t http://www.asktog.com/columns/038MacUITrends.html

6

Technical communication and accessibility

Klaus Schubert and
Franziska Heidrich

Abstract   Accessibility is a central notion in current societal debate. In this context, accessibility is often viewed primarily as an attempt at designing public buildings, means of transport, websites or printed documents in such a way that they can be used by people with disabilities. These are technical measures aimed at persons with mobility or perceptional disabilities. If successful, they allow the targeted group to perceive a certain piece of information. But what if the problem lies in understanding the contents? The present contribution, therefore, is concerned with linguistic and communicational measures intended to enable audiences with various kinds of abilities to perceive and understand information and to actively engage in communication.

Keywords   accessibility; optimizing communication; specialized communication; steered communication

Introduction

In this chapter, we review various types of barriers which impede people's passive or active participation in communication. Our focus lies on technical communication. However, neither the barriers nor the measures for overcoming or reducing them are limited to technical communication. Rather, there are interesting activities going on outside the sphere of technical communication proper from which our discipline can learn – and vice versa. This contribution therefore examines controlled languages, easy-to-read languages, plain languages and other means of deliberately steered communication. We focus mainly on linguistic and communicational measures steering the English language (if applicable) but sometimes we give a brief insight into the concepts of other linguistic communities.

We categorize occurring barriers to active participation in communication and, furthermore, take a first step in linking them with existing concepts for overcoming, reducing, or avoiding these barriers. The main questions are: which part of communication presents the main reason for the barrier in each case considered; on which part of communication does a measure mainly focus; and how are these two linked together.

Optimizing communication

One of the central ideas – as well as a fundamental feature – of specialized communication is the aim to optimize communication (Schubert, 2007, p. 366; 2011, p. 366). The basic ideas behind the optimization of specialized communication are the increased legibility, readability, comprehensibility and, deduced from those, usability of texts and other products of communication. The comprehensibility does not only address native, but also foreign-language recipients.

Optimization is a broad and multi-faceted activity. Its smallest unit is an act carried out by one or more actors. We call this act a controlling influence. Such an act is normally carried out intentionally. However, the effect the act has on communication may be a corollary of the goal the actor wishes to achieve, so that the actor may not even be aware of the communicative effect. In this sense, a controlling influence can also be exerted unintentionally.

Many controlling influences focus on optimizing the act of communication (Van Vaerenbergh and Schubert, 2010, p. 15). One main focus of the optimization of (technical) communication is thus to enable the recipient to access the required information (Göpferich, 1998b, p. 888; Schubert, 2006, p. 489). The most prominent aspects supporting that purpose are the comprehensibility, unambiguity and consistency of technical documents (Baumert, 1998/2010; Drewer and Ziegler, 2011, pp. 103 ff., pp. 108 ff.).

Controlling influences focus on all four dimensions of specialized communication (as suggested by Schubert, 2007, pp. 248–249) – the dimension of the (technical) content, the dimension of the linguistic form, the dimension of the technical medium and the dimension of the work processes. Controlling measures can be carried out on all of the four dimensions. The optimization in turn relates to the cognitive information processing of the concerned persons and draws on their knowledge (Strohner and Brose, 2002/2006, pp. 6–7). The success of optimizing measures depends on specific needs of the concerned actors and is above all useful in the fields in which communication is for a particular purpose (as it is the case with specialized communication) (Schubert, 2006, p. 488). The most sustainable intervention into communication is an intervention into the means of communication, as it does not only influence a single act of communication but determines the possibilities for every future use of that particular means of communication (Schubert, 2016, p. 17). The regarded means of optimization have one important aspect in common: the intervention into the means of communication happens intentionally.

The following section deals with the description of barriers to communication. It is then followed by the analysis of different simplified means of communication intended to overcome those barriers and rounded out by a first approach to systematize the aforementioned.

Barriers to communication

This section focuses on certain barriers persons with diverse communicative abilities are confronted with. The barriers primarily refer to the restraint of the active or passive participation in communication and include, but are not limited to, technical communication.

In order to approach a systematization of these barriers, as a first definition we consider those to impede certain (groups of) people from participating in communication (or certain fields of it) as successfully as other people. When talking of the existence of a particular barrier, the barrier should not be considered to be absolute. On the contrary, all of the barriers have an associated range of severity. For instance, when referring to a sensory disability, the hearing-impaired are not either completely deaf or not impaired at all – the level of impairment can be graded: *hearing, hard of hearing, extremely hard of hearing...* For simplification, we discuss all considered barriers as if they are absolute in the following but always imply a range of impairment.

In order to systematize the aforementioned barriers, we suggest a first categorization:

- Sensory barriers
- Domain-specific barriers
- Cultural barriers
- Cognitive barriers
- Language barriers
- LSP (Language for Specific Purposes)-related barriers.

Sensory barriers refer to the impairment of one of the human senses, for example the sense of hearing or sight. Persons who are visually impaired or hard of hearing are not able to perceive a message that is aimed at that particular sense.

Domain-specific barriers occur if a person does not have the technical knowledge (or not to a sufficient degree) that is necessary to understand a message. Technical communication (no matter the degree of specialization) always requires a certain amount of specialized knowledge (Kalverkämper, 1998 a, b). People confronted with a domain-specific barrier are indeed able to perceive a message, but fail to understand its content.

On the other hand, people confronted with a *cultural barrier* do not or not sufficiently or not as intended understand a message because they lack the necessary cultural prerequisites. According to a very generic definition of 'culture', we consider it to also include, for example, text type conventions concerning the linguistic form, the presentation and structure of contents or the graphic design of documents. In those cases, people might again perceive a message but do not understand it in part or entirely because they are not familiar with the textual design and thus cannot identify the intended speech act or illocution of the message.

A *cognitive barrier* affects people who do not understand a message because the conceptual structure of the message exceeds their cognitive abilities. Important aspects of those barriers are the linguistic and the content-wise complexity of a message. Again, the recipient does not lack the sensory prerequisites to perceive the message, but is nonetheless confronted with a barrier; in this case with a divergence between textual complexity and cognitive abilities. Aspects of linguistic complexity are, for example, the indirect representation of content, negation, or the grammatical conjunction of sentences or utterances.

The *language barrier* refers to messages that are composed of a language that the concerned person does not understand. This barrier does again not refer to perception but to comprehension.

The *LSP-related barrier* means that the concerned persons do actually understand the used language, but do not understand the relevant specialized language. This barrier can be linked to the domain-specific barrier, but not necessarily. Again, this barrier does not refer to perception but to comprehension.

It is quite clear that all the aforementioned barriers do not necessarily occur individually but can also occur combined with each other. People might be affected by several barriers of one type or by several barriers of different types. There are, for example, people who are affected by two of the sensory barriers, that means they are deaf and blind. Again, there are also recipients who are affected by various barriers of different types, which is, for example, the case if a deaf person wants to watch a foreign language movie. Even more complex situations are those in which one barrier leads to another one, as is the case for people who became deaf prelingually and thus do not perceive written language (which is derived from spoken language) as easily as the hearing. These people are confronted with a sensory barrier on the one hand and a cognitive barrier on the other hand.

Simplified means of communication

After the above approach of systematising barriers to communication, in the following section we regard different simplified means of communication which are intended to overcome, reduce or avoid these barriers in a certain way. We discuss linguistic and communicational measures that are intended to enable impeded recipients to perceive information in case of a perceptional barrier or to understand information in case of a comprehensive barrier, respectively, and thus to (re)gain the (partial) possibility of actively participating in communication and thus in society.

One main consideration before being able to discuss or even systematize different approaches for optimized communication is if the discussed barriers can be overcome, reduced or avoided. There is actually a basic difference between the three concepts, firstly with regard to the actor in charge. If we consider the barriers figuratively, 'overcoming' is something that the affected person has to carry out in person, while other persons are responsible for 'reducing' the respective barrier. If we take the nonsensory barriers as an example, the domain-specific, language, LSP-related and cultural barriers can likely be overcome by the person affected, through the acquisition of knowledge. That is also the case for cognitive barriers – but all the aforementioned with certain limitations. Not every person is able to acquire the necessary knowledge or skill to overcome certain barriers. In this case, it is necessary for other persons to reduce the barrier. This can be attained, for example, by the creation of more comprehensible messages which reduce the cognitive or language barrier, by the creation of additional explanations for the culturally or domain-specific impeded or by the extended use of common language instead of LSP for the LSP-related barriers.

Essentially, the sensory barriers cannot be overcome nor reduced by linguistic or communicative means. They can be avoided to some extent, though, by addressing other senses, as will be described in more detail below.

Easy-to-read

The easy-to-read concept is a standard for the English language which primarily focuses on people with intellectual disabilities and "can also be useful to make information easy for many other people to understand. For example, people who do not have English as a first language, people who find it difficult to read" (Inclusion Europe, 2009, p. 6). It contains rules on the word and sentence level and also refers to information structuring. Examples of the rules are:

- "Do not use difficult words. If you need to use difficult words, make sure you always explain them clearly." (ibid., p. 15)
- "Be careful when you use pronouns." (ibid., p. 15)
- "Keep your sentences short." (ibid., p. 17)

The easy-to-read concept is not based on a compelling set of rules, but rather describes some aspects of easy-to-read language in a very detailed way and leaves others out completely or just touches upon them. The rules are rather unspecific (for example concerning the exact length of an *easy* sentence) and the controlling influence of the easy-to-read concept is rather weak. It is meant to reduce cognitive and language barriers (Inclusion Europe, 2009, p. 6). Easy-to-read concepts gain more and more importance in different linguistic communities. For example, in Germany, the concept of 'Leichte Sprache' for people confronted with sensory, cognitive or language barriers has recently been regarded not only by representatives of the affected interest groups (cf. i.a. Netzwerk Leichte Sprache, 2015) but has also become a field of interest for an especially founded university research group, which has already resulted in much more detailed and principle-based linguistic rules (cf. i.a. Maaß, 2015, also cf. Baumert, 2016, on Easy-to-read concepts for the German language).

Plain English/plain language

Plain language is a concept intended to facilitate communication between the government, authorities, etc. and the people. It is implemented either by law (cf. i.a. Plain, 2011, for the USA), by means of a campaign by a commercial organization (cf. i.a. Cutts, 1995/2013; Plain English Campaign, 2016, for the UK) or by means of recommendations by, for example, universities, government authorities, etc. (cf. i.a. Bayerisches Staatsministerium des Inneren, 2005, for Germany). There is a wide divergence in terms of detail between the different nations (for example, a rather detailed regulation for the USA compared to rather diverse, short, undetailed recommendations for Germany).

Plain language is used by authorities or the government to communicate with citizens. As often technical topics are covered in those communicative acts (such as legal matters), plain language often contains elements of LSP.

Depending on the local situation, the controlling influence can be rather weak or pretty strong. It is, however, observable that most guidelines are more detailed than those for Easy-to-read. Plain Language aims at reducing cognitive, domain-specific, and LSP-related barriers.

Controlled language

Controlled languages are a means of communication for which the lexicon and the grammatical options are strictly reduced. They are designed predominantly for technical documentation. For the specification of a controlled language, a natural language is re-designed by means of reduction in that certain elements, relations, words, syntactic options, etc. are forbidden and, in part, replaced with others (Lehrndorfer, 1996 a, b; Göpferich, 1998 a; Schubert, 2001, p. 230, 2007, p. 334, 2008, p. 211; Crabbe 2010; Drewer and Ziegler, 2011).

Controlled languages intend to exclude the phenomena of a natural language which have a negative influence on the comprehensibility of a text, such as complexity, ambiguity, and redundancy (Ley, 2005, p. 27). Successful controlled languages are, to date, always linked with specific domains and/or text types (Drewer and Ziegler, 2011, p. 197). Controlled languages are specified in such detail that they can be supported by controlled language checkers, software which supports the author in keeping to the rules. The aims of controlled languages are the comprehensibility for native speakers, the comprehensibility for non-native speakers, and an increased translatability (Schubert, 2001). The most prominent and most detailed specification is *ASD Simplified Technical English*, which was originally designed for the European aerospace and defence industries (Crabbe, 2010, p. 56; ASD 2013; Simplified Technical English Maintenance Group, 2015). Similar concepts for other languages and other technical domains exist.

The controlling influence of controlled languages is strong, the regulations are compelling and detailed. Controlled languages aim at reducing cognitive as well as language and LSP-related barriers.

Planned languages

Planned languages are languages which have been consciously designed. In general, planned languages are not specified for a specific domain or for LSP, but rather for oral and written international and inter-ethnic communication (Back, 1996, p. 881). The most prominent planned language is Esperanto. A planned language is created by means of construction rather than reduction. It does not reduce the possibilities of a natural language but is mostly constructed from scratch (even though constructed languages lend some parts of the lexicon and grammatical principles from natural languages) (on the construction of planned languages cf. i.a.: Back, 1979, 1996, pp. 883–884; Blanke, 1985, pp. 88–98; Maxwell, 1989).

The controlling influence of constructed languages is strong. Constructed languages aim at the reduction of language and cultural barriers.

Sign language

Sign language is a concept for hearing-impaired people. Its main constituents are gestures, facial expression and posture. For those who are both deaf and blind, there exists a tactile sign language which is based on hand- or body-perceptible signs and gestures. As sign language is used in every area of life, from leisure to science, it also contains LSP. Sign language is a language-specific phenomenon, which means that there are different sign languages in different countries (cf. i.a. British Sign Language, Deutsche Gebärdensprache, Langue des Signes Française).

Sign language is very strictly defined and the controlling influence is very strong. Sign language aims at avoiding a sensory barrier.

Further areas for research

In this chapter, we try to propose a first systematization of communicative barriers and the simplified means of communication which aim at overcoming, reducing or avoiding these. The present chapter is just a quick glance into what has to be investigated. This chapter mostly focuses on simplified means of communication which are used in parts of the English-speaking world, while other languages are mostly excluded. Furthermore, the chapter mainly focuses (with some exceptions) on means of communication that are in some way or another situated in the sphere of technical/specialized communication and thus excludes other communicative fields. It is well known that simplified means of communication do not only exist for the English language, and that the methods of simplification may – and may even have to – apply to all areas of life.

This chapter only proposes a first step in analyzing and systematising the aforementioned barriers and means of overcoming them. An important project for the future would thus be to continue on this topic and to extend the focus. Other worthwhile aspects of analysis might for instance be a further systematization, a more detailed analysis of the width and depth of said measures, an analysis of to which extent people confronted with barriers can overcome or avoid them at all and to which extent the means of communication can assist them in doing so, etc.

Conclusion

The aim of this chapter was to propose a first systematization of said communicative barriers and the simplified means of communication to overcome them. The table below might give a more simple overview of the above.

Table 1: Communicative barriers and the respective means of communication

Means of communication	Barriers					
	Sense	Domain	Culture	Cognition	Language	LSP
Natural language	–	–	–	–	–	–
Easy-to-read	–	–	–	+	±	–
Plain language	–	+	–	+	–	+
Controlled language	–	–	–	±	±	–
Constructed language	–	–	+	–	±	±
Sign language	+	–	–	±	–	–

In the table, a minus refers to a barrier that is not avoided, overcome or reduced by the respective means of communication, while a plus refers to a means of communication that does reduce the respective barrier. A plus-minus, accordingly, means that the link exists to some extent.

Natural language is our baseline for comparison for all other means of communication. Easy-to-read primarily aims at the cognitive barrier and the use for language learners is only subordinate. Plain language has a focus on the increased comprehensibility of technical content, thus also focuses on the reduction of the LSP-related barrier. Controlled language has a minus in terms of domain-specific as well as LSP-specific barriers. That can be explained by the fact that controlled language does not reduce these barriers linguistically but by other means of optimization (for example, explanations). Constructed languages are intended to be intercultural but do only get a plus-minus in terms of language-related barriers as these languages also have to be learned.

References

ASD AeroSpace and Defence Industries Association of Europe (2013). *ASD-STE 100 Simplified Technical English. Specification ASD-STE 100. International Specification for the Preparation of Maintenance Documentation in a Controlled Language*. Issue 6. Brussels.

Back, O. (1979). Über Systemgüte, Funktionsadäquatheit und Schwierigkeit in Plansprachen und in ethnischen Sprachen. In H. Felber, F. Lang and G. Wersig (Ed.), *Terminologie als angewandte Sprachwissenschaft* (pp. 257-272). München et al.: Saur.

Back, O. (1996). Plansprachen. In H. Goebl, P. H. Nelde, Z. Starý and W. Wölck (Ed.), *Kontaktlinguistik / Contact Linguistics / Linguistique de contact.* (Handbücher zur Sprach- und Kommunikationswissenschaft 12.1.) (pp. 881-887). Berlin/New York: de Gruyter.

Baumert, A. (1998/2010). [pdf] *Gestaltungsrichtlinien. Style Guides planen, erstellen und pflegen.* First published at doculine Verlag. Hannover: SerWisS - Hochschulschriftenserver der Fachhochschule Hannover. Available at: http://serwiss.bib.hs-hannover.de/files/261/ Baumert_Gestaltungsrichtlinien_Style_Guides_1998.PDF [Accessed 06 Nov 2015].

Baumert, A. (2016). [pdf] Leichte Sprache – Einfache Sprache. Literaturrecherche, Interpretation, Entwicklung. Hannover: Hochschule Hannover. Available at: http://serwiss.bib.hs-hannover.de/frontdoor/deliver/index/docId/697/file/ ES.pdf [Accessed 22 Apr 2016].

Bayerisches Staatsministerium des Inneren (2005). [pdf] *Freundlich, korrekt und klar. Bürgernahe Sprache in der Verwaltung.* Available at: https://www.uni-augsburg.de/einrichtungen/gleichstellungsbeauftragte/downloads/ broschuere_freundlich_korrekt.pdf [Accessed 12 Dec 2016].

Blanke, D. (1985). *Internationale Plansprachen.* (Sammlung Akademie-Verlag Sprache 34) Berlin: Akademie-Verlag.

Crabbe, S. (2010). Controlled Languages for Technical Writing and Translation. In I. Kemble (Ed.), *The changing face of Translation.* Ninth Annual Portsmouth Translation Conference. Portsmouth, 7 November 2009 (pp. 48-62). Portsmouth: University of Portsmouth. Available at: http://www.port.ac.uk/media/contacts-and-departments/slas/events/tr09-crabbe.pdf [Accessed 15 Dec 2016].

Cutts, M. (1995/2013). *Oxford Guide to Plain English.* 4th ed. Oxford: Oxford University Press.

Drewer, P. and Ziegler, W. (2011). Technische Dokumentation. Übersetzungsgerechte Texterstellung und Content-Management. Würzburg: Vogel.

Göpferich, S. (1998a). Interkulturelles Technical Writing: Fachliches adressatengerecht vermitteln. Ein Lehr- und Arbeitsbuch (Forum für Fachsprachen-Forschung, 40). Tübingen: Narr.

Göpferich, S. (1998b). Möglichkeiten der Optimierung von Fachtexten. In L. Hoffmann, H. Kalverkämper and H. E. Wiegand (Ed.), *Fachsprachen/Languages for Special Purposes. Ein internationales Handbuch zur Fachsprachenforschung und Terminologiewissenschaft/An International Handbook of Special Languages and Terminology Research* (pp. 888-899). New York et al.: de Gruyter.

Inclusion Europe (2009). [pdf] *Information for all. European standards for making information easy to read and understand.* Available at: easy-to-read.eu/wp-content/uploads/2014/12/EN_Information_for_all.pdf [Accessed 12 Dec 2016].

Kalverkämper, H. (1998a). Fach und Fachwissen. In L. Hoffmann, H. Kalverkämper and H. E. Wiegand with C. Galinski and W. Hüllen (Ed.), *Fachsprachen/Languages for Special Purposes*. (Handbücher zur Sprach- und Kommunikationswissenschaft 14.1.) (pp. 1-24). Berlin/New York: de Gruyter.

Kalverkämper, H. (1998b). Rahmenbedingungen für die Fachkommunikation. In L. Hoffmann, H. Kalverkämper and H. E. Wiegand with C. Galinski and W. Hüllen (Ed.), *Fachsprachen/Languages for Special Purposes*. (Handbücher zur Sprach- und Kommunikationswissenschaft 14.1.) (pp. 24-47). Berlin/New York: de Gruyter.

Lehrndorfer, A. (1996a). Kontrollierte Sprache für die Technische Dokumentation. Ein Ansatz für das Deutsche. In H. P. Krings (Ed.), *Wissenschaftliche Grundlagen der technischen Kommunikation* (Forum für Fachsprachen-Forschung, 32) (pp. 339-368). Tübingen: Narr.

Lehrndorfer, A. (1996b). Kontrolliertes Deutsch. Linguistische und sprachpsychologische Leitlinien für eine (maschinell) kontrollierte Sprache in der Technischen Dokumentation (Tübinger Beiträge zur Linguistik, 145). Tübingen: Narr.

Ley, M. (2005). Kontrollierte Textstrukturen. Ein (linguistisches) Informationsmodell für die Technische Kommunikation. Dissertation. Justus-Liebig-Universität Gießen, Fachbereich Sprache, Literatur, Kultur. Available at: http://geb.uni-giessen.de/geb/volltexte/2006/2713/pdf/LeyMartin-2006-01-30.pdf [Accessed 10 Apr 2016].

Maaß, C. (2015). Leichte Sprache: das Regelbuch. Berlin: Lit.

Maxwell, D. (1989). Principles for Constructing Planned Languages. In K. Schubert, K. with D. Maxwell (Ed.), *Interlinguistics*. (Trends in Linguistics, Studies and Monographs 42.) (pp. 101-119). Berlin/New York: Mouton de Gruyter.

Netzwerk Leichte Sprache e. V. (2015). *Netzwerk Leichte Sprache*. Available at: http://leichtesprache.org/ [Accessed 15 Dec 2016].

Plain English Campaign (2016). *Plain English Campaign*. Available at: http://www.plainenglish.co.uk/ [Accessed 12 Dec 2016].

Plain Language Action and Information Network (PLAIN) (2011). [pdf] *Federal Plain Language Guidelines*. Available at: http://www.plainlanguage.gov/howto/guidelines/FederalPLGuidelines/FederalPLGuidelines.pdf [Accessed 12 Dec 2016].

Schubert, K. (2006). [pdf] Interlinguistik und Kommunikationsoptimierung. In A. Künnap, W. Lehfeldt and S. Nikolaevič Kuznecov (Ed.), *Mikrojazyki, jazyki, inter"jazyki* (pp. 486-492). Tartu: Universitas Tartuensis. Available at: http://www.klausschubert.de/material/Schubert2006InterlinguistikUndKommunikationsoptimierung.pdf [Accessed 10 Apr 16].

Schubert, K. (2007). *Wissen, Sprache, Medium*, Arbeit (Forum für Fachsprachen-Forschung, 76). Tübingen: Narr. Available at: http://d-nb.info/1045615382/34 [Accessed 15 Dec 2016]

Schubert, K. (2008). Konstruktion und Reduktion. In H. P. Krings and F. Mayer (Ed.), *Sprachenvielfalt im Kontext von Fachkommunikation, Übersetzung und Fremdsprachenunterricht* (pp. 209-219). Berlin: Frank & Timme.

Schubert, K. (2011). Optimierung als Kommunikationsziel: Bessere Sprachen. In K.-D. Baumann (Ed.), *Fach - Translat - Kultur. Interdisziplinäre Aspekte der vernetzten Vielfalt*. Vol. 1. (Forum für Fachsprachen-Forschung, 98) (pp. 363-392). Berlin: Frank & Timme.

Schubert, K. (2016). Barriereabbau durch optimierte Kommunikationsmittel: Versuch einer Systematisierung. In N. Mälzer (Ed.), *Barrierefreie Kommunikation - Perspektiven aus Theorie und Praxis* (Kommunikation - Partizipation - Inklusion 2.) (pp. 15-33). Berlin: Frank & Timme.

Simplified Technical English Maintenance Group (2015). [pdf] *ASD Simplified Technical English*. Brussels. Available at: http://www.asd-ste100.org/ [Accessed 25 Sep 2015].

Strohner, H. and Brose, R. (2002/2006). Kommunikation und ihre Optimierung. In H. Strohner and R. Brose (Ed.), *Kommunikationsoptimierung. Verständlicher - instruktiver - überzeugender* (Stauffenburg Linguistik, 30) (pp. 3-14). 2nd ed. Tübingen: Stauffenburg.

Van Vaerenbergh, L. and Schubert, K. (2010). [pdf] Options and Requirements. A Study of the External Process of Specialized Document Production. *Hermes* [online], (44), 9-24. Available at: http://download2.hermes.asb.dk/archive/download/Hermes-44-vaerenbergh&schubert.pdf [Accessed 10 Apr 16].

Part 2

Resources for Technical Communicators

7

Scalable video production for technical communicators

Jody Byrne

Abstract   The idea of using video to communicate technical information is nothing new. For many years, video has been used to great effect to provide technical training, software demonstrations and simulations. Recently, the increased availability and usability of video production tools, as well as the steady improvements in the infrastructure needed to deliver them, mean that increasing numbers of people and organizations are using video to explain how things work.

Even a cursory search through YouTube reveals countless how-to videos and tutorials on a vast range of technical subjects produced by professionals and enthusiasts alike. Clear evidence, if it were needed, that video content is no longer regarded as a high-tech novelty or as a 'nice to have' bonus for users. Instead, many users regard videos as their first port of call when they need help, and if manufacturers don't provide them, they'll find someone who will.

But with the proliferation of video as a means of communicating technical information come new challenges. While producing one or two videos on an ad hoc basis is quite straightforward, developing a model for scalable video production as part of an organization's user assistance (UA) strategy requires careful thought and planning. The fact that there are numerous 'fan' videos available on the Internet is a clear indication that many companies still have a long way to go in terms of producing the large volumes of instructional videos their users demand.

Keywords   video; accessibility; process; software; user assistance

Introduction

The aim of this chapter is to describe some of the key challenges involved in systematically producing high-quality how-to videos at a scale that meets users' expectations and needs. It is worth noting that the circumstances prevailing at any given company, combined with the nature of individual products and their users, mean it's not possible to provide definitive strategies on what you should do. Nevertheless, the following discussions – which are based on practical experiences of video production at both small companies and large multinationals – will help you to identify potential hurdles much earlier, and formulate your own solutions for overcoming them.

In this chapter we will examine issues such as:

- How the type of video you choose impacts your technical communication workflow
- Best practices for planning and producing how-to video projects
- Localization strategies: technical, organizational and linguistic issues,
- Improving the accessibility of videos for users with hearing difficulties
- Streamlining the video production process to take all of these issues into account.

Why use video at all?

For those of us who grew up in a world where you learned how to do something by *reading* instructions, videos might seem a little indulgent or unnecessary. After all, if you read the instructions enough times, you'll eventually figure out what you need to do. It may be tempting to dismiss video as a gimmick or something only for the young and the tech-savvy, but the reality is that video, in many cases, really is the best way to communicate technical information.

This is particularly true in the case of software, where graphical user interfaces – designed to minimize some of the complexity of using software – make the process of explaining processes using text alone much more complex. The task of writing instructions on how to navigate through an interface and perform various tasks is akin to giving someone driving directions over the telephone. The quest for precision and clarity – using just a single channel of communication – can result in exhaustive, and sometimes convoluted descriptions, of what users should see on-screen. Even then, there is no guarantee that users will interpret these instructions correctly.

While diagrams and screenshots go some way towards reducing the difficulty users may have in comprehending a visual process using just textual information, they lack the temporal context provided by showing processes in real-time. As such they can really only fill in *some* of the gaps in users' comprehension.

But even assuming that it were possible to perfectly describe everything a user needs to do in order to use a piece of software, the question arises as to whether this is the most efficient or indeed effective method of doing so. If it takes 4000 words and several screenshots to explain a process that, in reality, should take the user a couple of minutes to perform, does that represent efficient communication, and should we be content with taking up so much of our users' time?

Ultimately, we can summarize the benefits of video by saying it's faster and more efficient to *show* people what to do than it is to *tell* them what to do.

Getting started with video

What kind of videos do you need to produce?

Before looking at how to incorporate large-scale video production into your processes, you need to decide what kind of videos you want to produce. Although the basic medium is the same, the content, style and length of a video varies, depending on whether it is a tutorial, a training video, an explainer video or a how-to video. Table 1 on page 134 outlines the key differences between some of the main types of instructional video.

Animated GIFs as an alternative to video

Before we discuss the challenges involved in producing videos, it's worth noting that there is an alternative to the videos listed in Table 1.

A GIF is a graphics file (Graphics Interchange Format) which allows you to combine several images into a single file. In effect, this allows you to create an animation consisting of several frames, and which plays like a traditional video.

Table 1: Typical video formats used for technical communication

	Aim	Features	Length
How-to	Quickly and effectively convey procedural information necessary to perform a task using an application.	Screen recordings of the user interface form the main focus of how-to videos because the aim is to walk the user through a procedure or task. Recordings will be highly focused on explaining how to perform a single specific task with the result that actions will be precise, clear and performed at a moderate pace to ensure viewers can follow everything. Captions and call-out text may be used sparingly so as not to obstruct the interface. Although not absolutely essential in all cases, voiceovers are usually used to minimize the need for captions, which can distract the user or obscure actions being depicted on-screen. If you choose not to use a voiceover, it must be absolutely clear from the video or accompanying text what actions are being performed and why. Viewers may quickly lose interest in such videos so it is best to keep them as short as possible.	Ideally 2–3 minutes maximum. If procedures or tasks are more complex, they should ideally be split into multiple videos, dividing content at logical breakpoints.
Tutorial	Similar to how-to videos in that they provide procedural information, but they deal with longer, more complex process and provide explanations and analysis to explain the theory behind a process.	Screen recordings of the interface will form the main focus of this type of video but because an important aspect of a tutorial is discussion, producers may use the interface almost like a whiteboard, pointing at or highlighting screen elements, showing examples or demonstrating the process slowly at first with accompanying explanations, followed by faster repetitions, or transitions and time remapping to indicate the passage of time.	The duration can vary depending on the procedure but it is not unusual to see tutorials of at least 10 minutes.

Table 1: Typical video formats used for technical communication (continued)

	Aim	Features	Length
Overview	Provide a high-level description and summary of the main components and features of an application. They are produced for the purposes of familiarizing users with the interface and the general tasks which can be performed.	Also known as walkthrough videos, overview videos will typically start with on-screen text to list the features covered in the video before proceeding to show users around the interface and describing the key features along the way.	Generally 3–5 minutes in order to cover all of the main features of an application in high-level detail.
E-Learning	Provide in-depth knowledge about a process or technique, often requiring comprehension rather than simply conveying procedural information. E-learning videos generally form part of a larger programme of content and learning.	In-depth theoretical and procedural information with exposition, repetition and reflection. E-learning videos will contain a combination of on-screen text, diagrams and screen recordings. Interactive elements such as quizzes and clickable areas to reveal additional information are common.	Varies but, because individual videos form part of a larger suite of videos and as such not always capable of serving as standalone content, they are technically quite time-consuming.

The main differences are that they are stored as a common image file, and they can be displayed in a web browser without any specialized plugins. However, animated GIFs do not contain any audio, just visual information.

Since images need to be loaded completely before they are displayed, there are practical limitations on how big an animated GIF can be before they become a nuisance to users. If the files are very large the space where the image is displayed will either appear blank, or will contain a static image until the file downloads. This can cause frustration or even confusion for users because they may not realize that it is an animated image.

For this reason, animated GIFs need to be kept relatively small, and as such they are really only useful for very short animations, typically around 10 seconds. Compare this to the typical how-to video which should only last around 2–3 minutes. Nevertheless, they are a useful alternative for scenarios which require more than a single screenshot but which do not require a longer video.

The fact that they can generally be created using the same tools you would use for creating videos means they can easily be used to supplement videos.

Do you have the resources to handle video production?

An obvious question when embarking on any project is whether the necessary resources are available. But what exactly are these resources? Generally, to produce videos, you need four types of resource:

- Financial
- Technical
- Human
- Organizational.

Financial resources

Before you can make any decisions regarding if, and how, you are going to introduce video production to your UA strategy, it is worth assessing how much financial support is available. The level of budget which can be drawn on will significantly influence the choices you make with regard to the scale of video production but also with regard to the availability of technical, human and organizational resources. Financial support is needed in the first instance to buy the practical resources such as equipment, software and media assets like stock photos and footage, graphic design assets and so on. But financial support is also necessary to ensure staffing issues are addressed and that organizational issues, such as adjusting existing processes and handling any shifts in deadlines, are taken into account.

Technical resources

Workstations

Video files are big, they require lots of space for storage and need a relatively high-specification computer to produce them. For this reason, determining whether existing computer equipment is capable of handling the demands of video production is essential. Fortunately, how-to videos don't require the computing power needed to produce

Hollywood-style productions, but there are certain minimum requirements:

- Portability

  Laptops are preferable to desktop computers because it will be necessary to move to a quiet location to record voiceovers.

- Operating system

  There is a misconception that in order to do any multimedia work, you need to have a Mac. While this may have been true once, advances in both hardware and software mean that this is no longer the case, and many professional video producers use Windows machines. Indeed, in many working environments a Mac may pose problems in terms of general compatibility and the availability of software. If you are used to using Windows, there is no need to make a switch.

- Graphics cards

  Discrete graphics cards are better than on-board graphics because screen recording and video editing require a lot of memory to ensure smooth recording and fast rendering times. As a minimum requirement, you will need a graphics card with at least 2 GB of RAM.

- Internal storage

  When producing videos, it is advisable to have as much internal storage as possible to maintain the performance of your machine and to avoid running out of storage. Ideally, you should have two separate hard disks: one to host your applications and one to host your data so as to improve the speed with which videos can be rendered and to mitigate somewhat against hardware failure. Additionally, solid state drives (SSD) provide greater reliability against disk failures and improved speeds when working, in comparison with traditional hard drives with spinning disks.

- External storage

  Of course, even with a massive amount of internal storage on a production machine, it will be necessary to archive both the final videos and the project files used to create them at regular intervals. Aside from safeguarding files in the event of equipment failing or being lost, archiving files on an external storage device and also on a network drive frees up space and helps to keep your production machine running smoothly.

■ Monitors

To ensure suitable working conditions when performing such a visual activity, it stands to reason that monitors are an important part of the video production workstation. As a minimum, one large high-resolution monitor is essential. But the standard practice, especially among video producers, is to have two widescreen monitors to provide as large a working area as possible to reduce eye strain, improve usability and generally speed up the production process.

■ Audio equipment

Just as monitors are important for the visual aspects of video production, audio equipment is equally important. At the start of the process, a good microphone is essential in order to produce professional quality voiceovers. Although it may be tempting to use the same headset you use for making telephone calls, the sound quality is rarely good enough. Condenser microphones which connect to your laptop via a USB connection are widely available for as little as €40.

During the editing process, being able to accurately monitor the quality and volume of the voiceover is important to ensure professional results. Sophisticated headphones with dynamic sound enhancement features and noise cancelling may be perfect when listening to music, but when editing audio, headphones without any sound enhancing features are best so that you get an accurate indication of what the voiceover actually sounds like. Often referred to as *studio* or *monitoring* headphones, they are acoustically 'flat' and do not enhance the audio in any way. This helps you to get an accurate indication of what the audio will sound like, even on low-quality speakers.

Human resources

While focussing on the technical aspects of video production is understandable, especially if it is a new concept to an organization, the most important challenges relate to the human resources needed to produce videos.

Who will produce the videos?

An important factor in determining how to implement video production as part of your UA strategy is identifying who will actually be responsible for producing the videos and whether they have the necessary skills needed to do it. Video production is a time-consuming process and this will have serious implications for lone technical communicators. If there is one technical communicator for a product, it is not reasonable to expect a whole series of videos on top of the existing UA output. To address this, it is useful to examine whether videos can be produced *instead of*, rather than *in addition to* certain deliverables, so that the technical communicator's workload does not become unmanageable. If there are several technical communicators available, it might be feasible for one person to be exclusively responsible for video production so as to provide consistency and to shoulder the additional demands. Depending on the volume of videos that is required, there may be a case for recruiting dedicated video-production staff.

Who will produce the voiceovers?

When you produce how-to videos, not only do you have to make sure that your script is clear and easy to understand, you also need to ensure that it is communicated in a clear and easy to understand way as well. Although voiceovers aren't essential for how-to videos, they make for a more engaging and effective experience for users, but only if they are produced properly. A video with a voiceover which has bad sound quality, or where the speaker's diction and pronunciation makes it difficult to understand what is said, can have catastrophic effects not only on comprehension but on the reputation of your company.

If you have a clear speaking voice and can communicate clearly to tell a compelling story, providing the voiceover yourself can help speed up the production process. However, you may not want to do this or, for branding reasons, it may be desirable to use someone with a particular accent, in which case your choice of voice talent will be determined by:

- Their ability to speak in a pleasant, clear and easy to understand manner

- Your company's branding guidelines; for example, all official videos must feature US or UK English, with no regional dialects

- Their ability to communicate enthusiastically and not sound dull or bored

- Availability and access to recording equipment.

Organizational processes

The issue of whether a company's work processes can comfortably accommodate the additional demands of video production is something that can only be answered by those familiar with those processes. We will discuss the impact of video production on a technical communicator's typical workflow later on, but an initial examination of the company's development processes and schedules should be carried as soon as possible to see if there are clear schedules and timelines which will facilitate the timely production of videos to the required standard. Inflexible processes or overly aggressive deadlines can hinder video production to the point where technical communicators find it impossible to produce videos on time without sacrificing quality.

Choosing the right software

The tools you will ultimately use to produce videos would seem to be the most important consideration when making the move to wide-scale video production, but the reality is that it's actually quite far down the list. Nevertheless, your choice of tool and what you expect it to be able to do can have implications for your choices later on, so it's worth discussing it now.

Producing videos for software products involves the following tasks:

- Creating screen recordings
- Editing screen recordings
- Recording and editing audio for voiceovers
- Combining recordings with other assets such as images, screenshots, text, audio, and other video footage
- Producing the final file in an industry-standard format.

Video tools

To minimize the learning curve for video producers and to reduce the complexity of the process, it is advisable to choose a tool that provides as much of the above functionality as possible in a single application. For example, a tool should ideally allow you to create screen recordings and then edit them without the need to import the footage into a new application. Similarly, an editing tool needs to allow you to perform basic audio editing tasks. Two of the more popular tools which can do this are *Camtasia* by TechSmith, and *FlashBack* by Blueberry Software.

When creating screen recordings, it is advisable to opt for a tool which allows you to record full motion. Certain tools – such as *Adobe Captivate*, frequently used in the production of e-learning materials – work on the basis of slides and are not fully aimed at capturing full-motion recordings, but rather provide this functionality as a limited additional feature. In many projects this is not a problem, but it can be problematic in the case of modern interfaces which often contain sophisticated animations because of the difficulty in capturing and editing the animations smoothly. As a rule, you should check that any screen recording tool is capable of full-motion recording.

Audio tools

As we mentioned earlier, the ideal video-production tool will provide as many of the necessary tools and functions as possible to simplify the workflow and to minimize the cost. The tools mentioned earlier all provide audio functions to allow you to record and edit your voiceovers. They also provide simple effects such as volume control, noise reduction and transitions.

However, there are standalone audio tools which can be useful, even if you have an all-in-one tool like *FlashBack* or *Camtasia*. *Audacity* is a free, open-source audio tool which allows you to record, edit and process your voiceovers. When combined with the *LAME* plugin to save audio in mp3 format, its clear interface and range of functions, including a powerful noise-removal effect make it a useful part of your toolkit.

Production considerations

How video affects your UA process

The typical UA process varies from company to company, and even from team to team but, at its most basic, it might involve the following stages:

1 Identify topics or functions that need to be documented
2 Conduct research with subject matter experts
3 Plan, structure and write draft
4 Secure clean test system to produce screenshots
5 Create and edit screenshots
6 Submit for approval or QA
7 Implement edits

Adding video as one of your outputs will inevitably involve changes to your process. If we look at the typical steps involved in creating a how-to video, we can get a sense of what these changes might be.

1. Identify the topic which requires a video
2. Conduct research with subject matter experts
3. Write and refine script and storyboard
4. Get approval for script
5. Locate voiceover artist and record voiceover
6. Postproduction for voiceover
7. Secure clean test system with which to produce screen recordings
8. Edit screen recordings and voiceover
9. Produce the final video
10. Submit for approval or QA
11. Implement edits by repeating steps 5–9 as necessary
12. Upload to server

As anyone who has had to replace countless screenshots as a result of seemingly minor UI changes will know, it's always best to leave the screenshots to as near the end of a development cycle or sprint as possible to minimize the likelihood of having to redo the screenshots. This is even more important with videos because it's not just a case of replacing a screenshot: you have to recreate all of your screen recordings, make sure they fit the timing of your video and voiceover and then render the video again. If you can't reach an agreement with the product owner or project lead to produce your UA to a schedule that gives you sufficient time between the end of development and release to customer (for example, some technical communicators work one sprint behind the main development team) you will need to either curtail the scope of your videos; that is, only produce videos for certain, more stable, topics or produce different kinds of videos, such as product overview videos or walkthroughs.

Additional steps for video

The key challenge when producing videos is that the preparation and production stages can take much longer than for standard documentation. This is because additional work is required and because more people may be involved.

Scripting

It might be tempting to say that there is no need to write a script for a video; after all, you've written a user guide, so surely you can just record that. Right? In fact, what you write in a voiceover script will be significantly different to standard documentation in terms of style and content. The basic distinction between writing something to be read as opposed to writing something to be spoken means that you'll need to make some substantial edits to ensure the voiceover flows, and that there are no unintentional tongue twisters. You also need to make sure that the voice talent can actually read a sentence comfortably and without running out of oxygen.

Similarly, you won't need to include many of the verbal cues for users, such as indicating the location of something on the interface, for example. The beauty of video is that you can easily show people what they need to do, so you don't need to duplicate your efforts by telling them as well.

Approval for scripts

Getting approval for your video script may not be something that you have to do in your company but it is highly advisable to do it anyway to avoid time-consuming and potentially costly edits later on, or in some cases having to completely re-work the entire video. This is particularly important when it comes to the voiceover, because even minor revisions to the script mean the entire voiceover has to be re-recorded. This might sound extreme, but the tone and quality of a person's voice can vary substantially depending on the time of day, the room in which they do their recording or on their general health and mood. With this in mind, it's virtually impossible to record the new lines of script and then paste them into your voiceover without listeners immediately noticing it.

The most useful people to enlist for the purposes of approving scripts are product owners or lead developers who can check your script for accuracy before you start work. You'll need to schedule this to take place much earlier in the process than usual before you can arrange to have the voiceover recorded. It is advisable to write this step into your UA schedule so that your approvers know that this is a recurring task and they will need to set time aside during each release cycle.

Voiceovers

Voiceovers take time to prepare, record and edit, and you need to factor in additional time to do this. If you are producing the voiceover yourself, you can save some time but you will need to clear some time in your schedule to find somewhere quiet with good acoustics where you can produce your recording without interruptions. If you are asking someone else – either colleagues or external voice talent – to produce the voiceover for you, you need to make sure that they are available when you need them so let them know as early on as possible that there is an upcoming project.

Storyboards

Producing a high-quality how-to video requires more than just a good script. With videos, you are communicating through two separate communication channels: auditory and visual. To make the most of this dual-channel means of communication, and to avoid potential confusion for viewers, you need to produce a storyboard which carefully charts the course of the video, mapping what is said with what is shown on screen at any given time. The format of a storyboard can be quite simple, but it is vital that you produce one so that you can better identify what resources you will need (for example, images, screen shots and screen recordings) and make arrangements for getting them on time. Your storyboard can also serve as a shopping list for all the assets you will need in order to complete your video.

Table 2: Sample storyboard

#	Voiceover	Visuals	Instructions
1	To create a new profile, click on Profile and select the type of profile you want to create.	Show main window and open Profile dialog box.	Highlight (zoom) the profile types; fade out.
2	Next, specify where you want to save your profile.	Show save dialog and add text caption [Network Drive]	Wait until the bottom folder is opened before displaying caption.

Test systems

Having a clean test system with realistic data is an essential requirement for anyone tasked with documenting software. Customers don't want to see screenshots of software containing text along the lines of 'Test1234' or 'This is a sample entry'. Nor do they want to see a list of Star Wars characters, or worse still, the names of real customers in the interface. With screenshots, it's possible to fix many of the common errors like real user names or error codes using a graphics editing tool. With video, however, it's not so easy. Editing individual parts of a screen recording is not always possible. Even when it is, it can require specialist software like *Adobe After Effects* and can be extremely time-consuming. Since video displays the system in real time, everything needs to be working perfectly: the system needs to respond quickly and predictably, there can't be any error messages or random glitches. All of this means that having a reliable test system with clean data is even more important when producing videos. This creates a certain amount of tension between the video producer and developers: video production needs to start sooner rather than later but test systems are often only available later on in the development cycle.

Accessibility

Although the combined use of audio and visual information means video is highly effective at communicating technical information to the majority of users, it is problematic for a large section of society who have some degree of hearing loss. It is estimated that some 360 million people – or 5% of the world's population – have what can be classified as "disabling deafness" (World Health Organization, 2015). Although there is some disagreement in the literature, due mainly to varying definitions of hearing loss, it is estimated that hearing loss affects 15% of people in the USA, and 11.2% of people in France (National Institute on Deafness and Other Communication Disorders, 2016; Kervasdoué & Hartmann 2016). By virtue of the fact that it replaces written information with the spoken word, video effectively excludes such users because the information being communicated is no longer accessible to them.

One relatively easy solution is to provide subtitles as a way of representing most of the audio information which would otherwise be lost. Tools such as *Camtasia* allow you to add closed-caption subtitles to videos. Creating them is simply a matter of editing the voiceover to condense it and then pasting it into subtitles. Admittedly, subtitles are far from perfect as they can distract from the on-screen visuals or even obscure them. As an alternative, providing a transcript can also help, especially if it is linked to the video so that the relevant sentence is highlighted in time with the video, as is the case with tutorials on Lynda.com.

Translation issues

In many jurisdictions, the European Union being a prime example, there are various legal provisions which require manufacturers to provide user documentation in the language of the end user. This means that it is it likely that the videos you produce will need to be localized. This requires discussion with management to determine which languages are required. This is largely determined by the overall localization and marketing strategy for your product and should be decided upon by the product owner or senior management.

One solution would be to localize all videos only for your most important markets, with selective localization of a small subset of videos for all other markets. This presupposes that videos are not the only source of UA for your users and that conventional documentation exists so that users can learn how to use the software. Indeed, this is vital to ensure that all legal requirements regarding the provision of user documentation are adhered to.

Once the issue of which languages videos should be localized into has been addressed, a decision needs to be made as to what extent videos are to be localized. Several options are available, each with its advantages and disadvantages:

Table 3: Video localization options

Localization option	Description
Full Localization	Fully localized video with translated voiceover and UI recordings. This is the 'gold standard' for video localization because it provides an identical experience for target users. However, it is expensive and just as time- and labour-intensive as producing the original videos, and as such will incur additional costs and increase the turnaround time.
Partial Localization	Translated voiceover with UI recordings in the original language. This is a compromise approach and may be suitable for bilingual countries or in countries where the original language is widely spoken. From an instructional design perspective, the tension caused by having to process two languages simultaneously may cause fatigue and confusion for viewers, and ultimately limit the effectiveness of the video. There is serious potential for confusion if, for example, functions and menu options are referred to using different words in different languages.

Table 3: Video localization options (continued)

Localization option	Description
Text-to-Speech (TTS)	TTS involves using text-to-speech software to produce a translated voiceover. The voiceover script is first translated and then processed by TTS software. This approach is faster and cheaper than using voice actors but the synthetic, artificial voice may irritate or alienate viewers, so its use should be considered very carefully.
Subtitles	This approach simply involves adding foreign language (or interlingual) subtitles to the original video. From a practical viewpoint, this is the fastest and cheapest option, but is problematic in how-to videos for a variety of reasons, not least that it distracts users from the interface being shown on screen, it can cause confusion because of different ways of referring to functions etc. and the subtitles themselves can obstruct important on-screen information.

Volume and frequency of releases

The decision of how many videos to produce and how frequently they should be delivered depends partly on the staff resources available and on the localization issues which were discussed earlier.

If you are producing videos which only need to be localized into a small number of languages and there is sufficient capacity to produce them, it may be possible to produce several how-to videos and an overview video for each release. If staffing is an issue, or where videos need to be localized into many languages, it may only be possible to produce a small number of how-to videos for the main features for each major release. There may be cases where it is only possible to provide an introductory overview video for each language. Collaborating with your customers can provide invaluable insights into which features and functionality you should cover in your videos.

Publishing

Publishing videos has the potential to cause problems if discussions about expectations, requirements and processes do not take place in advance. In order for users to be able to watch the videos, the videos need to be stored on a server somewhere. Such a server needs to be reliable, have sufficient bandwidth to allow large numbers of users to watch videos without interruption, and it must provide enough storage space to store all of your videos. There are many options available, each with its own advantages and disadvantages.

Hosting large numbers of videos on the same server as your website is technically possible but not advisable unless you have sufficient capacity, bandwidth and storage. Large volumes of users requesting videos at the same time may effectively choke the server, causing your website to respond very slowly or even go offline.

Services such as YouTube, in addition to being free, offer the kind of massive infrastructure that can deal with millions of users and large volumes of data. Indeed, many companies use YouTube to host promotional videos, but its use for how-to videos should be approached with caution because it may not fit with your company's corporate image, branding or commercial strategy. More importantly, however, is the fact that if you ever need to update a video, you cannot simply replace the video; you must upload the new version as a new video. This new version will have a new URL, which you will need to update in all associated documentation and web pages.

Other platforms such as Vimeo or Kaltura provide reliable infrastructure with powerful management tools, including the ability to update videos while maintaining their URLs, but such services are not free and can be expensive, depending on your requirements,

Roll-out

Once the decision has been made to proceed with large-scale video production, and all of the resources have been identified and put in place, there is still a certain amount of work that needs to be done before you can roll out this concept. This work can help improve acceptance and facilitate the smooth adoption of video production.

Selling the idea

Having spent a lot of time researching the benefits of videos as a way of communicating with users, it can be easy to assume that the benefits are so obvious that they don't need to be explained. It might be tempting to wonder why anyone in their right mind would even question the value of videos. But for many of the stakeholders who will be affected, these benefits may not be quite so clear. Indeed, the potential changes may cause sufficient concern that the promise of efficient, clear and modern communication is simply not enough to provide whatever enthusiastic reassurance you might give.

The people you need to convince as part of the video rollout process include:

- Product managers, who may be concerned that videos may not provide the kind of detailed technical information they think their users need;

- Development colleagues, who are concerned that the changes you want to make to processes, and demands for clean test systems will adversely affect their work;

- Technical communicators, who will have to produce these videos and who may be concerned about their workload or even job security, particularly if they don't have a background in digital media. This is perhaps the most important group to cater for.

The challenge here is to be open, honest and very clear about what will be involved. Focussing on the benefits alone will not be enough to convince some stakeholders, so it is important to show that you acknowledge some of the concerns and potential problems and then explain what measures you have put in place to counteract them.

For the technical communicators who will ultimately be responsible for producing videos and who are concerned about the increased workload or the need to learn new skills, you can provide details of the training and resources that you will provide, such as cheat sheets, templates, technical support etc. It might also be useful to communicate changes in UA policy that might say, for example, that certain documents can be replaced by videos, or provide details of the number of videos that need to be produced per release.

Training

Designing a suitable programme of training and support for video producers is absolutely essential in order to succeed with the rollout. Colleagues with little or no experience in this field will need reassurance that videos are not particularly difficult to produce once they have the right training and resources. Even colleagues who have previous experience should attend the training to make sure that everyone is following the same, standardized process. The challenge of scalable video production with multiple producers is making sure that the same standards are applied and that all videos have the same look and feel, no matter who made them. The training can be reinforced by providing all video producers with a set of style guidelines and specifications for videos. Another issue to consider is that, even if colleagues take a training course, it may be some time before they have to put it into practice, by which time they have forgotten much of what they have learned.

Providing on-going support and advice on technical and design issues as well as short how-to videos is crucial to ensure continued and problem-free support for video production.

Visual identity and branding

The need for consistency can also be addressed by creating standard templates that conform to your company's corporate branding guidelines. Templates and a collection of standard assets, such as icons, also reduce the need for individual producers to create their own assets or obtain assets from different sources, and as such, help to create a common visual identity across the various teams and products. The decision as to how prescriptive templates should be is something that will depend on your company's culture, branding guidelines and also on the diversity of products that you make.

Conclusion

The use of videos is undoubtedly a useful addition to the range of tools used by technical communicators. For users, it can make the process of learning how to use software more engaging, more efficient and less demanding. For some technical communicators it provides an exciting new dimension to conventional technical communication and is something which they will embrace enthusiastically. For others, however, it is completely uncharted territory and may be a cause of anxiety due to the perceived complexity of the process and the unfamiliarity of the technologies involved.

While the process of actually creating a video is not particularly demanding from a technical perspective, there are various practical and organizational issues which need to be addressed in advance in order to facilitate the large-scale production of video content. Issues such as defining a strategy for producing videos, identifying and gathering the necessary resources and generating support from stakeholders are critical for ensuring scalability in the context of a particular company.

Once these issues have been addressed, the remainder of the process centres on the skills of individual technical communicators. Perhaps somewhat surprising is the fact that most of the skills required to produce a how-to video are already core parts of a good technical communicator's skillset:

- Research and analytical skills
- Communication and interviewing skills
- Time management and scheduling
- Information design and writing skills
- Basic graphic design skills, gained, for example, from creating screenshots.

The only additional skills that are required involve using audio tools and video editing tools. These are essentially mechanical tasks that can be learned and performed quickly and with a minimum of training.

With this in mind, scalable video production is a matter of preparation and planning, and it can be achieved more easily than you might expect.

References

Kervasdoué, Jean de & Laurence Hartmann (2016) *Economic Impact of Hearing Loss in France and Developed Countries: A survey of academic literature 2005–2015.* European Hearing Instrument Manufacturers Association [online] Available at: http://www.ehima.com/wp-content/uploads/2016/05/FinalReportHearingLossV5.pdf [Accessed 31 Jan 2017].

National Institute on Deafness and Other Communication Disorders (2016) *Quick Statistics About Hearing*, [online] Available at: https://www.nidcd.nih.gov/health/statistics/quick-statistics-hearing [Accessed 31 Jan 2017].

World Health Organization (2015) *Deafness and Hearing Loss, Factsheet No. 300*, [online] Available at: http://www.who.int/mediacentre/factsheets/fs300/en/ [Accessed 31 Jan 2017].

8

Managing digital complexity in technical communication

Marie Girard and
Patricia Minacori

Abstract

While technical communication teams are already busy managing growing volumes of content, new challenges arise. With the sheer diversity in authors of technical content, publication platforms, and formats available, technical communication projects tend to grow out of control. A new, more systemic approach is required to address complexity in technical communication.

In this chapter, we first discuss the origins and principles of systems thinking, and describe why this approach can help address the challenges of the growing complexity technical communicators are facing today.

Then, we describe the practical advantages of systems thinking for auditing, governing, and planning content:

- Organizations are usually not prepared for the level of flexibility required for all-online, all-connected communication. Systems thinking embraces feedback loops and hidden interactions by fostering interactions between stakeholders, regardless of their place in the organization chart.

- In a volatile environment, managing content audits and models cannot be done in a monolithic way. Systems thinking correlates and compares various models as facets of a bigger complex reality.

- While short-term planning remains feasible, long-term planning becomes impossible in an uncertain business context, where disruption can happen at any time. Systems thinking prioritizes strategy over long-term planning, so that every individual in the organization understands the long-term goal and can quickly make decisions towards that goal at any time.

We conclude that a systems thinking approach can complement the more standard analytical approach, adding flexibility and agility to the necessary structure of content systems.

Keywords   technical communication; complexity; systems thinking; content strategy; change management

Introduction

Technical communication is becoming more and more complex because of changes in terms of volume, veracity, variety, and velocity of content.

Volume

As companies transition from paper-based information to online formats, the amount of online content grows exponentially (SINTEF, 2013). On top of existing content that needs to be maintained, new content gets created.

Veracity

Gone are the days when technical publication teams were the only ones to create technical content. Marketing teams have realized the value of credible technical content as a means to drive sales. Support teams have found a way to promote self-help and to reduce costs by making their knowledge base content available online. Technical experts publish content as a means of promoting their expertise. In fact, social media now enables anybody to publish technical content. The multiplication of authors and sources makes it more difficult to assess the authenticity and accuracy of a given piece of information.

Variety

With the digitization of products and services, content is everywhere: web interfaces, mobile applications, and connected objects. We now face the limits of 'write once, publish everywhere' with content that needs to be broken down into smaller and smaller units in order to morph into this growing variety of outlets.

Velocity

Widespread use of social media now requires interactivity and more dynamic means of delivery, such as video, infographics, blogs, and forums. Technical communicators must be able to adjust their content to these quickly evolving needs.

Most innovations in technical communication so far have consisted in defining structures and tools to better manage content: its creation, storage, categorization, and conversion into multiple formats. While these activities are undeniably useful, they are based on an analytical approach, and they reach their limits when faced with the complexity of content today. So while the technical communication teams are busy making their content management systems more efficient, business changes affect the customer experience more and more strongly, as Guenther (2013, p. 18) describes it:

> When dealing with enterprises, we are used to strange and often quite frustrated experiences. They seem to make even simple transactions awkward and complex. Straightforward activities such as booking tickets for a journey, paying your taxes, subscribing to a health insurance, or resolving a problem with your energy supplier require customers to embark on a laborious journey, jumping between call centers, online forms, and missing information. They make us shift between different contacts, tools, and communication channels. They lose track of the conversation, get stuck in inflexible procedures, and often fail to deliver what they promised. Most of us have had a lot of such experiences, be it with companies, government institutions, or other types of enterprises, making them appear slow, rude, and inhumane.

The complexity of our business context has consequences on the customer experience, and we are seeing the limits of analytical approaches to our content issues. Systems thinking can help address such cases when complexity makes it too difficult to manage change.

Origins and principles of systems thinking

Western scientific thought has its foundations in rationalist thought, which has descended to the present in a direct line from Greco-Roman antiquity, and particularly from Aristotle.

In France, René Descartes laid down four foundational principles in his *Discourse on the Method* (1637):

- The first principle is never to accept anything as true that could not incontrovertibly be known to be so; that is to say, carefully avoid both prejudice and premature conclusions; and include nothing in judgements other than what presents itself so clearly and distinctly that there would be no occasion to doubt it.

- The second principle is to divide all the difficulties under examination into as many parts as possible, and as many as are required to solve them in the best way.

- The third principle is to conduct one's thoughts in a given order, beginning with the simplest and most easily understood objects, and gradually ascending, as it were step by step, to the knowledge of the most complex; and positing an order even on those which do not have a natural order of precedence.

- The last principle is to undertake such complete enumerations and such general surveys that one would be sure to have left nothing out (Descartes, 2006, p. 19).

From that time until the start of the 20th century, any scientific discipline had to include the following conditions:

- Definition by its subject
- Description of an objective reality
- Explanation of this reality by applying a method of investigation based on observation, measurement and the repetition of facts.

These methods were suited to the study of stable systems, constituted of a limited number of elements, which it would be possible to identify by using reason and analysis, so as then to explain the causes and functioning of these systems (Comte, 2002).

However, in the 1930s, many scientific thinkers no longer believed themselves capable of explaining, or even of acting, in the face of the world's complexity by proceeding from analytical thought (Bériot, 2006). Indeed, not all the principles laid down by Descartes have positive consequences. Lapointe (1993) highlights that:

- Knowledge has been fragmented into as many areas as there are phenomena capable of being studied, and the fact of focussing on one thing at a time does not allow us to deduce which of its characteristics belong to the whole

- Disciplines have become isolated from each other
- They have become overly specialized
- Communication between specialists has become increasingly difficult.

According to Durand (2013), three movements are capable of providing answers to the complexity of the world: structuralism, cybernetics, and information theory. It would also be possible to include constructivism in this list.

Structuralism

This movement, which emerged at the beginning of the 20th century, is particularly important in the humanities and social sciences, especially in linguistics (Saussure, Hjelmslev, and Jakobson), anthropology (Levi-Strauss) and psychology (Gestalt theory, Piaget). It is based on the postulation that objects in the world are not capable of being known in themselves, but through the relations of their elements between themselves and with the whole of which they are a part. For example, in linguistics, the notion of structure applies at every level of linguistic analysis; that is, one can speak of morphological, syntactic and semantic structure. The notion of structure proceeds from the idea of a system that functions according to laws, and which is preserved or enriched through the interplay of these laws without anything being added from the exterior. In this sense, the notion of structure in linguistics can resemble a closed system. It is possible in the case of linguistics to see the merit of such a system of thought, which endeavours to recognize the elements that can be interrelated within a system. But it is also clear that a closed system does not consider the relationships that it can have with other systems. If, in the framework of structuralism, a text can only be expressed by its reading alone, one might understand henceforth that the author of the text and the person to whom it is addressed (to mention them alone) are of fundamental importance in the understanding of a text. Thus, structuralist linguistics has given way to theories of enunciation, for example, in which the subject takes on particular importance.

Cybernetics

This term comes from the Greek *kerbeineiké*, deriving from *kuberman*, which means 'to direct' or 'to govern'. Cybernetics is linked to the art of steering. In 1950, Norbert Wiener (1961) enunciated a science of self-regulating systems, a science whose study would allow for the control, and the placing into communication, of man, animal and machine, because he saw humans and machines as pure organisms of relating and communicating. In this framework, Wiener's approach consisted of recognizing the structure of the interior of a machine or of an animal, then in describing its relationship and interactions with its environment and, finally, in predicting its behaviour and evolution over time. He chose to focus, in particular, on the notion of 'feedback' which he defined as a response relationship from a target element back to a source one.

Wiener (1961) highlighted two types of 'feedback':

- Positive feedback, in which a reaction goes in the same sense as a principle action, thus creating a 'snowball' type of action, or a cumulative effect
- Negative feedback, in which the size of an input action is compensated for, and which assures that this input is regulated, as is the case of a thermostat on a boiler.

This notion of feedback, which now seems to be a part of our daily lives, represented a true innovation at the time.

Cybernetics is the science of information and regulation. Its objective is both the knowledge and the steering of systems. Numerous scientific sectors have evolved from these applications, particularly the automating of production, computerized networks of communication, and new management methods for organizations.

The Macy Conferences of the mid-20th century were the starting point for this discipline. In fact, in 1942 and between 1946 and 1953, scientists, engineers, philosophers, neurologists, anthropologists, psychologists and economists reflected on newly emerging ideas about the brain. Nine conferences were organized in New York and a tenth in New Jersey, under the auspices of the Macy Foundation. According to Dupuy (1994), the work of these conferences would go on to establish cognitive sciences.

Information theory

In 1949, an engineer at an American telephony company, Claude Shannon, and a philosopher, Warren Weaver, published a work about the transmission of information and the theory of communication which would have a world-wide influence in many domains. At this time, Shannon's intention was to obtain the best output from a physical transmission system between a transmitter and a receiver, in order to keep production costs to a minimum. This theory stated that an information source produces a message. This would be encoded by a transmitter which would despatch the message along a channel towards a receiver. The message would then be decoded by the receiver, which would have to reconstruct it from the basis of a signal, in order to transmit it to a destination. Finally, this message could be mixed up or deformed by a factor called noise.

This model of communication did not take into account the elements that made up information – it did not include the notion of a feedback loop. It is thus considered as static. However, it came to have a large impact and its first and foremost function came to be the study of cryptography.

This model was enriched, in particular, by Harold Lasswell (1948) who described in details the different elements that made up information. This American sociologist was interested in the importance of a means of communication for disseminating political ideas and propaganda. His project was to respond to the following questions: Who says what? In which channel? To whom? With what effect? By doing so, he was taking back up the questions posed by Quintilian (c. A.D. 35–100). The principal research lines of enquiry would evolve around the description of transmitters, the analysis of content, the study of channels of transmission, the identification of receivers, and the evaluation of the effects of communication. These elements of analysis are echoed in the daily business concerns of technical communicators.

Communication between groups expanded greatly in the 20th century with the birth of new information and communication technologies (ICT). According to Durand (2013), the 19th century was the energy century: it involved replacing human and animal energy by mechanical energy. But the 20th century was that of communication involving new means of transport, new media and especially new technologies such as computers, cell phones and the Internet. This communication revolution was accompanied by a process of expanding complexity – greater and greater amounts of increasingly complex data started to be exchanged throughout the world and in businesses.

In the 1950s in the United States, thanks to the contributions of structuralism, cybernetics and communications theory, General System Theory emerged, formulated by Ludwig Van Bertalanffy (1968), which would then become systemics. He spoke of the emergence of open systems which would exchange matter, energy and information with their environment.

All systemics experts recognize that it is difficult to find a single definition of the theory. For Van Bertalanffy, systems are "sets of elements standing together in interaction" (1968, p. 38). He saw 'closed systems' as "systems which are considered to be isolated from their environment", and stated that "a system can be defined as a complex of interacting elements" (1968, pp. 55).

Checkland explained that a system is: "A model of a whole entity; when applied to human activities, the model is characterized fundamentally in terms of hierarchical structure, emergent properties, communication and control" (1981, pp. 317–318).

According to Bunge, "A system may be said to have a definite composition, a definite environment, and a definite structure. The composition of a system is a set of its components; the environment, the set of items with which it is connected; the structure, the relations among its components, as well as among these and the environment. For example, a theory is composed of propositions or statements. Its environment is the body of knowledge to which it belongs (for example, algebra or ecology): and its internal structure is the entailment or logical consequence relation" (1979, pp. 4–5).

Last but not least, the French Society of Systemics proposed a definition in 2003: "Whatever is too complex to be reduced must have recourse to systemics, a new discipline which includes theoretical, practical and methodological approaches. The latter considers problems in relation to observation, representation, modelling and simulation. It also specifies borders, internal and external relations, structures, laws or emerging properties." (Donnadieu et al., pp. 2–3.)

The question is then to determine the fundamental concepts that govern systems thinking. Durand (2013) lists four concepts: interaction, wholeness, organization and complexity.

Interaction

This is a fundamental and rich concept. Interaction is not necessarily a cause to effect action between two elements. Rather, it can take complex forms. For example, in the science of communication, Bateson (2002) distinguishes four levels of interaction:

- Purely visual interaction
- Language and communication of ideas

- Imitation
- Suggestion.

Another example of interaction is illustrated by 'feedback', which was initiated by cybernetics and which has fundamental importance in numerous domains ranging from training to change management.

Wholeness

If, according to Van Bertalanffy (1968), a system is composed of elements, it cannot be reduced to the sum of its parts. This idea had already been demonstrated by Blaise Pascal in his Pensées (published 1662): "I hold it to be equally impossible to know the parts without knowing the whole, and to know the whole without having a particular knowledge of each part" (1995, pp. 33).

The system is, therefore, more than a whole or aggregate form. It is especially in a relationship with its environment. Therefore, it has emergent qualities that its parts do not possess. From this, one can talk about a hierarchy in systems, and the more one ascends in the hierarchy, the more complex the systems become.

Organization

This is a central concept in systemics. It is a process, which may be illustrated by the example of information being laid out in such as way as to produce a documentation plan. In this framework, organization implies an objective of optimizing the components of systems and their arrangement. This organization can take a structural or a functional aspect.

The structural aspect highlights four components:

- A border that separates the system from its environment and which is more or less permeable
- Elements that can be identified and sorted
- A network of relations, with regard to communication and information
- Reservoirs to allow the system to adjust correctly.

The functional aspect can be studied in the form of:

- Flows, for example, information flows
- Decision centres, which receive information and turn it into actions

- Feedback loops, which allow for solutions to be tested out, and for the decision-makers to be informed
- Response times, which allow adjustments to occur within the pre-arranged time-frames.

In this functional aspect, systems can be broken down into sub-systems, which one can also call modules. This break-down can prove to be useful in the case of the Agile method, for example, which, in the domain of information technology (IT) project management, has the objective of delivering modules incrementally. The Agile method came into being in 2001 through the determination of specialists from the world of software. They thus created a 'Manifesto for Agile Software Development' (http//agilemanifesto.org/iso/en.principles.html). For "designing better ways of developing software" the Agile method gives priority to:

- Individuals and interactions over processes and tools
- Working software over comprehensive documentation
- Customer collaboration over contract negotiation
- Responding to change over following a plan.

The sub-modules can also be broken down into hierarchical structures.

Complexity

Complexity is inherent to every system, particularly in businesses that are confronted with an exponential increase in information and in products. Edgar Morin dedicated six volumes of his *Method* to this field of study (1977, 1986, 1999, 1994, and 2005). The root of the term comes from the Latin *complexus*, which means 'woven together'. The complexity of a subject is linked to the number of its characteristics and especially to the links between its elements. It must also confront uncertainty, the risks of the environment, and it must respond to the ambiguous relations between order and disorder.

One must not confuse complication with complexity. A complicated phenomenon can be resolved if one gives the necessary time, energy and money to it.

Edgar Morin (2005) defines complexity as a phenomenon that seems first of all to be quantitative, that is, involving an extreme quantity of interactions and interferences between a very large number of units. It also includes uncertainties, indeterminate elements, and random phenomena. It always involves chance. Complexity also refers to uncertainty within richly organized systems. It is a mix of order and disorder. Here we are also speaking about the black box principle, which cybernetics and communication theory had indeed recognized, but which they avoided without refuting it.

Phenomena are complex because they require multi-dimensional knowledge. In order to try to delimit the multi-dimensional character of a phenomenon, it is futile to study its different elements, but it is useful rather to consider the relations between its elements and the holistic character of an organization. This "complexity arises from a great number of interactions and interferences between a great number of units" (Morin, 2005, pp. 92). It is vain to want to know everything, or to have total knowledge. It becomes impossible to reach a state of completeness.

Edgar Morin listed four principles that can illustrate complexity:

- The dialogical relationship, which is about linking several or different logics that may appear to be both contradictory and complementary at the same time.
- Recursive organization, as is shown by feedback loops. Products and effects are both cause and producer, at the same time, of what is produced. This is the case of tornados, which are at the same time cause and effect.
- The hologrammic principle, according to which the whole and the parts are connected. Morin used the principle of the hologram in which its least part contains almost the totality of the information. This principle allows one to consider that the whole cannot be reduced to the sum of its parts, because in a complex system, it is of foremost importance to take into consideration the interactions between the parts.
- The meta point of view principle linking the observers, the objects observed and their environments.

Hologrammic organizations are rich because:

- The parts may be singular or original while still possessing the general or generic characters of the organization of the whole.
- The parts can be endowed with relative autonomy.
- They can establish communication among themselves and carry out organizational exchanges.
- They can even be capable of regenerating the whole (Morin, 2015).

According to Morin, it is necessary in a context of complexity to put into operation cognitive strategies whose goal is to:

- Extract information from the 'ocean of noise'
- Produce correct representations of the situation
- Formulate possibilities and scenarios for action.

Starting from these fundamental concepts, Durand (2013) goes on to explore the issues with which all systems are confronted:

- They have to master their relationship with the environment. This is characteristic of open systems that engage in exchanges with the outside world. They adapt to changes but must maintain a balance and not lose their energy, information, or content.
- Closed systems are those that refer back entirely to themselves, or in other words, which are entirely self-contained.
- The notion of interface comes from this first issue. An interface is an area of exchange between distinct systems or between sub-modules. In certain organizations the role of interface is essential: it is a true pivot for feeding the system. Technical communicators play this role of pivot in organizations.
- Systems are organized into modules (functional organization) and levels (hierarchical organization).
- They have to ensure their survival.
- They need variety. They possess states and configurations in order to adapt to challenges. But to steer these systems, it is also important to establish coordination so as to avoid chaos.
- Finally, systems evolve in order to respond to the needs and demands of organizations, markets and products. They adapt to their environment, develop, and innovate. This is at the heart of the issues involved in change management.

The complexity of the world has caused the principles of systemics to be back on trend, even though they were developed in the 1950s. A holistic – no longer a solely analytical – approach has allowed us to develop a larger area of focus and to attempt to understand and act on another level. Systemics has allowed us to progress in an understanding of the end goals that are sought when dealing with every difficulty, because systems thinking helps modelling problems. Simon (1950) speaks of not looking to establish an absolute rationality, but to consider a limited rationality. Within this framework, it seems important to try to reason at a reasonable level, and therefore to set out a study at a judicious level, that enables us to progress in the resolution of difficulties more quickly or to circumvent complexity by putting in place the best and richest principles of regulation and adjustment possible.

Systemic approaches to design and development

Dominated by rational, linear, and analytical thinking until the Internet revolution, design and development practices have now evolved into more systemic ways of dealing with the complexity brought by connected networks of data.

In what Senge (2006, p. 271) calls the "Knowledge Age", as opposed to the "Industrial Age", previous mechanistic approaches give way to more organic means of addressing problems, where the focus is on the whole rather than the individual parts, on interactions between individuals, and on systems that include the observer.

Two design and development methods have emerged recently in the software industry that reflect this change of mindset: Design Thinking and the Agile method.

- Design Thinking is a problem-solving approach to design. This method steers away from analyzing problems as a means to find solutions. Instead, the focus is on collaboration and ideation to come up with a solution.

- Agile is an approach to development that was created to help software teams adapt to fast-changing business conditions. Agile does not reject traditional ways of managing projects (processes, tools, comprehensive documentation, contract negotiation, and following a plan). It shifts the focus to activities that make it easier to deal with complexity: interactions between individuals, collaboration with customers, prototyping and reactivity to change.

Design Thinking and Agile are used more and more in different types of industries, but their scope remains limited when it comes to content.

Indeed, when dealing with content creation, these approaches are not enough to embrace the full scope of a 'content ecosystem'. Content creation goes beyond software or product, and has implications in culture and collaboration. As Morville (2014, p. 26) puts it: "When we see everything through the lens of software and startups, we lose our peripheral vision. Information systems aren't just code. They are also about content and culture." Content is more than product code and interfaces; it requires a wider type of collaboration that encompasses the whole content ecosystem.

According to Senge (2006, p. 270), "The problem starts with not understanding knowledge, how it is created and how it operates in practical settings – because knowledge is social." Therefore, "to manage knowledge you need to address collaboration and tools to help collaborate." Content has a double role of internal collaboration and external communication (Guenther 2013, p. 166). Because content is based on language, it is inherently tied to human interaction.

Managing complexity in technical communication implies a change of scope, beyond production of product content to include collaboration between teams. Because information has become ubiquitous, and because information is essential in the workings of systems, the scope of technical communication management needs to encompass both means of internal collaboration across silos and general business strategy. In the information age, technical communication draws closer to change management.

Managing complex content systems

Our analytical cultural bias could lead us to think that change can be brought to complex systems by analyzing them, identifying levers of change, and then pushing these levers. But complexity makes this far from obvious.

To manage change in complex content systems, a different approach is required, based on observation from within the system, collaboration, experimentation, and letting go of control.

According to Rockley (2003, p. 450), "Starting with a thorough analysis is key to a successful unified content strategy. Thorough analysis ensures that your strategy addresses your organization's specific needs and goals." This approach reaches its limits when the complexity of the content ecosystem makes it impossible to lead a thorough analysis.

These limits are encountered when analysis becomes faced with what Senge (2006, p. 57) called The Laws of the Fifth Discipline:

1 Today's problems come from yesterday's 'solutions'.
2 The harder you push, the harder the system pushes back.

3　Behavior grows better before it grows worse.

4　The easy way out usually leads back in.

5　The cure can be worse than the disease.

6　Faster is slower.

7　Cause and effect are not closely related in time and space.

8　Small changes can produce big results – but the areas of highest leverage are often the least obvious.

9　You can have your cake and eat it too – but not all at once.

10　Dividing an elephant in half does not produce two small elephants.

11　There is no blame.

These laws may seem completely illogical if you follow a linear and analytical approach. And if you apply analytical methods to drive change in a complex content ecosystem, you are likely to push the change in the wrong direction (Meadows 2008, p. 145).

This level of complexity is often referred to as 'wicked problems'. These types of problems "require holistic, cross-domain thinking beyond the conventional analytic and decision-centric mind set." (Guenther 2013, p. 57.)

Leverage points are not as easily accessible as in more simple systems. As Meadows puts it (2008, p. 165) "There are no cheap tickets to mastery. You have to work at it ... throwing yourself into the humility of Not Knowing."

Here are some recommendations for technical communicators who work with complex content systems and want to improve customer experience in this context: create multidisciplinary teams, leverage collective intelligence to manage and audit content, innovate with experimental transversal projects, and let go of centralized control on content creation.

Creating multidisciplinary content teams

Systems thinking focuses on interactions. Many different groups are now contributing to the publication of online technical content: marketing, support, technical experts... These groups usually work in what are often called 'silos'. Quite often propositions for managing a better content experience involve 'bridging' or even 'smashing' these organizational silos.

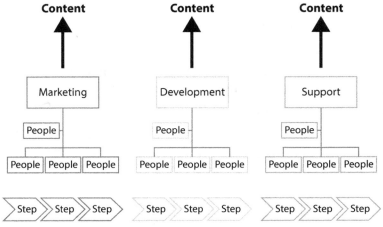

Figure 1: Organizational and content silos

A systems-thinking approach to the silos issue consists of observing interactions within the system, to start with (Meadows, 2008). By looking beyond the silos, one can see the nature of interactions and the feedback loops that happen naturally within the system. You can do this by interviewing stakeholders of the content ecosystem.

So before making any attempt at coordinating across silos, a technical communicator willing to tackle the content problem in a holistic way will interview the various people in the organization who produce content.

Figure 2: Stakeholder interview sheet

The second step is to organize content creators into a group, not with a view to controlling their actions, but with a view to fostering interaction between these stakeholders. This has a very concrete, direct effect on the resources used to create content: people from entirely different teams start helping each other out in creating content. Ideas for content reuse get raised. New feedback loops emerge in a very natural, organic way. As Appelo (2012, p. 5) notes, "when you mix different ideas from multiple sources, a new idea can emerge that both aggregates and improves on the pre-existing ideas." It is important to find a good mix of business, technical, and content experts in the group so that each can bring their unique point of view to the table (Welchman, 2015).

Scott Page (2010) demonstrated the value of diverse groups in addressing complex problems, and stated that groups that are made up of many people who think in different ways can trump groups of people who are very bright but alike. Involve expertise and methods from different professional backgrounds, including those with business-, technology-, and people-centric perspectives.

Leveraging collective intelligence to manage and audit content

One of the attributes of complex systems is that they behave like a black box, because of the quantity of elements and their relations within the system. Completeness becomes impossible to reach.

Common approaches to content management suggest starting with an inventory. With content volume growing exponentially, making such an inventory becomes a task that requires so much energy that it takes all the bandwidth and does not enable analysis. Models and storage systems have become a moving target because of the sheer volume of content they cover.

The same goes for metrics. By attempting to measure content ROI (return on investment) and effectiveness through clicks and bounce rates, we treat content as a thing while it should really be treated as a living organism, where function matters more than the individual parts.

Moreover, the quantitative measures we can retrieve about content tell us little about its quality, and even if they did, because everything is interrelated, it is not possible to define a clear cause–effect relationship between a measure and what to do to improve the content based on that measure.

According to Morville (2014, p. 90), "our numbers tell us what not why; they calculate the future as the sum of its past; and shape how we think and what we do more than we know." In other words, the real value one can gain from quantitative metrics comes from the qualitative conversations that can be held about these metrics. For Meadows (2008), if quantity is the measure, quantity will be the result. Qualitative metrics are harder to define and measure, but they remain essential for shaping a good strategy.

In a complex environment, collective intelligence can work wonders in identifying areas of leverage for improving content. Quantitative measures, inventories and classifications can become the basis for group analysis, provided the group is diverse enough to shed light on the various facets of a given problem.

If it's impossible to have a single monolithic model and quantifiable means of measurement, having several of those is still better than none, and can prove really useful when you see them as facets of a more complex reality: that is, multidimensional knowledge. Leadership needs to understand requirements of digital content experience (qualitative). Digital content workers need to understand business objectives and performance (quantitative).

To leverage collective knowledge, bring together the various categories and metrics you have for content, and ask stakeholders to talk through them, while enabling other stakeholders to comment freely. Leave space for a qualitative assessment of the content experience. That's the value of conversation.

Capture ideas for improvement that come out of the discussion, and validate these ideas with the group.

Innovating with experimental transverse projects

With the support of collective knowledge of areas that require attention, you can start acting on these improvement points. We mentioned earlier that the boundaries of the content system were very large, spanning beyond products or documents, which is the reason for its complexity.

By narrowing the focus to a specific problem, you can use a design thinking and/or Agile approach to developing a solution; that is, you can collaborate iteratively on a prototype. Use the feedback principle of Agile to check the effects of the prototype on the system, and adjust your course of action based on this feedback.

You can then establish innovation teams across organizational boundaries, fostering a culture of shared problem ownership.

Experimentation is key, as Appelo (2012, p. 4) points out: "We need to understand how to influence the whole system by poking at it. Then we see how it responds. As change agents we try to nudge people, teams and organizations so that they will reorganize themselves."

For experimentation to be successful in content ecosystems, the content itself must be flexible enough to be reused, repurposed, and republished. Semantic tagging and structured content make this possible.

Letting go of control of content creation

The business context has become more and more unpredictable. The rise of social media, mobile, and connected objects blurs the lines between high-tech and more traditional industries, and all businesses are rushing to adjust to these shifts in technology.

Short-term planning remains possible:

- Support creates content based on customer issues
- Doc fixes inaccurate content based on feedback
- Marketing organizes campaigns.

With the current unpredictability of business, however, long-term planning becomes impossible. The only way to enable a fast-enough response to change is to enable teams to make decisions quickly and autonomously. So that anyone, anywhere in the organization, can create content that aligns with the system's purpose. This type of decision-making model is biomimetic, as Senge (2006, p. 269) highlights: "We know that in healthy living systems, like the human body or a wetlands, control is distributed."

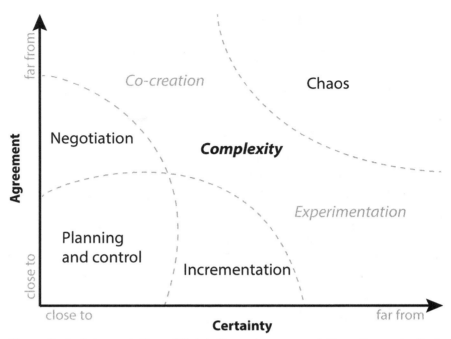

Figure 3: An interpretation of Ralph Stacey's representation of complexity in management

It becomes, therefore, all the more important to clarify and communicate the purpose of an organization's content, so that all content creators can make decisions about what to create and when, without having to follow a predefined plan. Use an enterprise-wide design approach to provide tangible visions of a meaningful, viable, and feasible future.

While push for control might easily disturb a self-organized group, a shared goal can help overcome the problems raised by power struggles (Meadows, 2008).

This involves shifting the focus from control and plans to building a shared understanding of the organization's purpose, and how content serves that purpose. This is radically different from the idea of creating a strategy from the top that will trickle down to the lower levels of hierarchy.

Here, the strategy resides in building empathy for customers, and surveying content creators to understand where they see the usefulness of their content: what are the core values behind the content? Let feedback not only apply to how values get applied, but to the values themselves.

A long-term strategy should focus on expected user experience: what will they decide to do after using your content, and how will they feel?

Guenther (2013, p.166) summarizes the importance of user experience as follows:

> To have an impact on the daily business, content needs to be defined first and foremost in terms of the meaning it creates for its audience. Meaning is inferred not from the content itself in isolation, but also from the context wherein it is perceived by people. That context includes what the content is about, who the originator and recipients are, as well as the chosen medium and structures to communicate it.

Sense-making, within the organization and outside, is becoming key to shaping a successful user experience.

Projections for the future

Faced with today's complex content challenges, there is a rising need for actors who are part of systems but can also understand these systems and have an approach to change that embraces complexity.

Technical communicators can play a role in the digital transformation of organizations to face more and more complex problems.

Content is about language and communication between humans: this is what makes it inherently complex. Clarity of information is key to making systems work. Technical communicators master this language of clarity. For Meadows (2008, p. 175) states, "The first step in respecting language is keeping it as concrete, meaningful, and truthful as possible – part of the job of keeping information streams clear."

Indeed, "most of what goes wrong in systems goes wrong because of biased, late, or missing information." (Meadows, 2008, p. 174.)

Technical communicators have a long history of playing a mediator role between experts and users. They understand the needs of stakeholders and mitigate these with user needs. This puts them in the right position to shift the focus on collaboration and empathy. They can take their connector role further to translate between domains, bridge viewpoints, coordinate efforts, and synthesize holistic approaches.

Guenther (2013, p. 53) calls this "hybrid thinking":

> Hybrid Thinking expands on the ideas of Design and Systems Thinking, both as the starting point for a challenge and as a focal point of any outcome. It is based on empathetic, intuitive thinking to create meaningful human-centric experiences. It follows the idea that the approach and thinking of great designers apply to any problem, even to those outside the traditional realm of design, to things like processes, business models, or enterprise transformation. When combined and integrated with other ways of approaching a problem, especially those applied in Systems Thinking to deal with complexity, [hybrid thinking] can be used to explore the problem space and generate possible outcomes.

Conclusion

Understanding complexity is becoming a key skill of technical communicators. Because of the nature of content, content ecosystems encompass much more than products and documents. With digital transformation, the lines between physical and information products start to blur.

Holistic approaches based on systems thinking can help manage the level of change caused by digital transformation. These approaches, however, require a definitive change of mindset to solve problems. They assume that instead of looking for quick levers of action, you start looking at how people interact before you induce change, and let interactions within diverse groups of people lead to collectively defined solutions. Instead of controlling content creation, you instil a shared sense of purpose to enable decision-making at any level of the organization.

Technical communicators who embrace the complexity of content ecosystems will thrive applying their hybrid thinking within and outside the organization, and will leverage their sense-making skills to create impactful user experiences through content.

References

Appelo, J. (2012). *How to Change the World: Change Management 3.0*. Rotterdam, The Netherlands: Jojo Ventures BV.

Bateson, G. (2002). *Mind and Nature: A Necessary Unity (Advances in Systems Theory, Complexity, and the Human Sciences)*. NJ, USA: Hampton Press.

Bériot, D. (2006). *Manager par l'approche systémique*. Paris : Eyrolles.

Bunge, M. (1979). *Treatise on Basic Philosophy, Volume 4. Ontology II A World of Systems*. Dordrecht, Netherlands: D. Reidel.

Checkland, P. (1990). *Soft Systems Methodology in Action*. London: John Wiley & Son.

Descartes, R. (2006). *A Discourse on the Method*, Ian Maclean Translator, New York: Oxford University Press Inc.

Donnadieu G., Durand D., Neel D., Nune, E. & Saint-Paul, L. (2003). *L'Approche systémique: de quoi s'agit-il ?* Synthèse des travaux du Groupe AFSCET "Diffusion de la pensée systémique". Available at: http://www.afscet.asso.fr/SystemicApproach.pdf [Accessed 3 Feb 2017].

Dupuy, J-P. (1994). *Aux origines des sciences cognitives*. Paris: Editions la Découverte.

Durand, D. (2013). *La systémique*. Collection " Que sais-je ", Paris: Presses Universitaires de France.

Guenther, M. (2013). *Intersection: How Enterprise Design Bridges the Gap between Business, Technology, and People*. Waltham, USA: Morgan Kaufmann.

Lapointe, J. (1993). *L'approche systémique et la technologie de l'éducation*, EDUCATECHNOLOGIES, Vol 1, No. 1. Available at: http://www.sites.fse.ulaval.ca/reveduc/html/vol1/vol1_no1.html [Accessed 3 Feb 2017].

Lasswell, H. (1948). "The Structure and Function of Communication in Society." *The Communication of Ideas. Ed. Bryson, Lymon. New York: Institute for Religious and Social Studies*, pp. 37–51.

Meadows, D. (2008). *Thinking in Systems: A Primer*. White River Junction, VT, USA: Chelsea Green Publishing.

Morin, E. (1977). *La Nature de la nature*. Paris: Seuil.

Morin, E. (1986). *La Connaissance de la connaissance*. Paris: Seuil.

Morin, E. (1990). *Science avec conscience*. Paris: Seuil.

Morin, E. (1994). *La complexité humaine*. Paris: Flammarion.

Morin, E. (1999). *Les Idées*. Paris: Seuil.

Morin, E. (2005). *Introduction à la pensée complexe*. Paris: Seuil.

Morville, P. (2014). *Intertwingled: Information Changes Everything*. Ann Arbor, Michigan, USA: Semantic Studios.

Page, S. (2010). *Diversity and Complexity*. Princeton, New Jersey, USA: Princeton University Press.

Pascal, B. (1995). *Pensées*, Penguin Classics.

Senge, P. (2006). *The Fifth Discipline: The Art and Practice of the Learning Organization*. London, UK: Random House Business Books.

Simon, H.A. (1960). *The New Science of Management Decisions*. New York: Harper & Row

SINTEF (2013, May 22). Big Data, for better or worse: 90% of world's data generated over last two years. *ScienceDaily*. Available at: www.sciencedaily.com/releases/2013/05/130522085217.htm [Accessed 3 Feb 2017].

Stacey, R. D. (1999). *Strategic management and organisational dynamics: the challenge of complexity*. New York: Financial Times/Prentice Hall.

Rockley, A. Kostur, P. & Manning, S. (2003). *Managing Enterprise Content: A Unified Content Strategy*. Indianapolis, Indiana, USA: New Riders.

Von Berlalanffy, L. (1968). *General System Theory, Foundations, Development, Applications*. George Braziller: New York.

Welchman, L. (2015). *Managing Chaos: Digital Governance by Design*. USA: Rosenfeld Media LLC.

Wiener, N. (1961). *Cybernetics or Control and Communication in the Animal and The Machine*. Second edition, Cambridge, Massachusetts: The MIT Press.

9

Writing good API documentation: an expert's guide, by a complete beginner

Neal Goldsmith

Abstract

This is a case study charting the author's journey from never previously having written any API (application programming interface) documentation to launching a well-received software developer hub. It's intended to provide the reader with the background, understanding and inspiration needed to tackle their own API documentation project, with current best practices in mind. Through the author's learning process, the discussion considers common problems with API documentation, the difficulties that can be encountered when trying to improve documentation, how to overcome those difficulties and ultimately attempts to identify some central tenets of good API documentation.

Keywords

API; documentation; problems; resources; feedback

Introduction

APIs are becoming increasingly important and prevalent. "To some degree, not having a public API today is like not having a website in the late 1990s", IBM's Claus T Jensen (2015, p. 1) states in the introduction to his *APIs for Dummies*. The title of a recent book on building APIs by Phil Sturgeon (2015) provides further acknowledgement of their universal ascendancy – *Build APIs You Won't Hate: Everyone and their dog wants an API, so you should probably learn how to build one*. In our interconnected world in which applications, services, devices, wearables and even household things are progressively communicating and integrating with each other, APIs are the underlying key driving this capability forward.

The likelihood is that you've already encountered an API in your normal line of work. Even if you haven't, you've no doubt made use of one, whether you've realized it or not – for instance when using a food delivery app on your mobile phone to order a pizza or when receiving a calendar notification that you're due to check-in for a flight.

Businesses are fast harnessing the money-making potential of APIs to expose services and make data available to external audiences. This enables integration and creation of new revenue streams. For some companies the API *is* the product, such as the online payment solutions provider Stripe and the cloud communications platform service Twilio.

As expectations and demands surrounding APIs change, so too do the demands and expectations surrounding API documentation.

But API documentation is not easy to write; or, more accurately, it's not easy to get it right. An API has no user interface, unlike a software application. It's not intuitive in the way an application can be. An application has a look and a feel and a layout, with visible prompts, entry fields and buttons to click. In this respect, the documentation is *everything* for a developer or anyone else coming to an API for the very first time. The documentation *is* the interface. Take that a step further: the documentation *is* the API, and in the case of Stripe and Twilio, it is in fact the *product* too. Once a business comprehends this, the importance of good API documentation becomes undeniable.

Documentation can make or break the take up and adoption of an API. Developers will talk about the documentation's quality – or lack thereof – and a reputation will be born that goes before it. No matter how good and well-designed an actual API is in itself, if developers can't begin to get to grips with it because of poor documentation, then they're not going to persist with it.

If a user interface is over-complicated and offers a poor experience, people won't use it and will go elsewhere. The same applies to API documentation.

How does one go about writing good API documentation, then? Where does one begin? How does one begin? Is anyone or anything out there explaining exactly how any of this is best done?

These were the questions I asked myself when required to write the revamped documentation for the second version of my company's API. In fact, that's not quite true. The first question I really asked was, "How does one write API documentation at all, let alone *good* API documentation?" I had never done it before.

When we launched our developer hub, though, the new documentation was a success. However, it took some time to get there.

This is my story then: an honest examination of the resources, tools and overall approach I used.

I'll cover what was inadequate and problematic with our existing API documentation. I'll also look at the obstacles that needed to be overcome in trying to improve the documentation, and how they were overcome.

Through my journey from beginner to 'expert' (and my tongue is lodged firmly in my cheek there), I hope to provide a solid grounding in API documentation to anyone similarly new to it, like I was. I also hope to provide inspiration to anyone looking to revamp existing content. I'll outline what I believe to be best practice throughout and will conclude by identifying what makes up the key components and characteristics of good API documentation.

An extremely brief history of APIs

APIs have been around for a while. They're not new – but they are evolving.

To loosely define an API, it allows two software systems to communicate and exchange data and actions with each other, via a set of methods and operations. These methods and operations have parameters which define the type of information and data that gets exchanged. A developer will use an API's methods and operations to make a request to an API. The API will respond and the developer can use the response and the data provided in programming their software system.

APIs were used long before the web. They were similar in basic principal to the web APIs of today, in the sense that they had documented and accessible entry points which allowed a system to interact with another system. However, their practice was a more closed and internal convention in comparison to current usage.

With the advent of the web, SOAP (Simple Object Access Protocol) and REST (REpresentation State Transfer) soon arrived and the age of publicly available web APIs was born.

SOAP was created in 1998 as a specification for exchanging structured information between web services in computer networks. However, while being flexible – it can be used with any transport protocol, including HTTP – it gained a reputation for being too verbose and thus complex to build and to use. This is largely because a SOAP message is an XML document containing various elements, such as an envelope (identifying it as a SOAP message), a header (containing header information) and the body (containing call and response information).

REST was then conceived and defined in 2000 by Roy Fielding in his doctoral dissertation, answering a desire to provide a universal way for software to communicate. Its popularity and usage has outstripped that of SOAP. REST is an accessible architectural style directly using the HTTP protocol, providing greater simplicity, performance and scalability. REST APIs are resource-based, which means the API calls a resource on a server via an endpoint (which is much like a web address, for instance https://my.api.com/v1/this-is-a-resource). The calls use HTTP verbs – such as GET, PUT, POST and DELETE – to create, retrieve, update or delete the resource (otherwise known as the acronym CRUD, encapsulating the four key functions of data storage).

REST has several appealing advantages over SOAP:

- It has a uniform interface; endpoints are always called, HTTP verbs are always used and calls always receive a response with an HTTP status.
- It is stateless; each request is completely self-contained – the server doesn't need information or influence from anything outside of the request in order to process it.
- There are clear operational boundaries between the two systems communicating; one server is always making the requests and the other is always being called.
- Responses can be cached; the server doesn't need to repeatedly and expensively go beyond the caching level to retrieve the previously requested information.

RESTful web APIs' accessibility and readability make them easier for developers to comprehend and to use, making them (theoretically) easier to document too.

It's no coincidence that businesses we recognize as Internet giants were among the first to realize the potential of web APIs. These companies were pioneers in launching them as part of their offering: Salesforce and eBay in 2000, Amazon in 2002, Flickr in 2004, then Facebook and Twitter in 2006. By and large, public web APIs are RESTful and they're continuing to mature and proliferate.

Web APIs have served as the catalyst for e-commerce and social networking. They've given rise to the cloud and they're currently propelling advanced mobile development. Next they look set to galvanize the Internet of Things.

Whereas a business only used to be able to reach an online audience and online consumers through their website, APIs are serving to enable potentially limitless outreach.

The situation I was in

I started as the new technical writer at my company, a UK-based software-as-a-service (SaaS) email marketing automation platform, at the beginning of 2013. Despite having worked within IT and software services since 2000, becoming a dedicated technical writer represented a career move and a completely new role for me.

The company was growing fast. They had a second version of their web API version that was coming out of a successful beta phase, available in both REST and SOAP.

The API can be used by anyone who has got API user credentials, which can be created by platform account owners. Using these credentials, the API can then be used to connect with almost any system. Data can be kept in sync between the two systems and can be imported and exported on a schedule, while most common tasks available in the system can also be automated.

As such, internal and external users consume the API for various reasons. It powers the key e-commerce and CRM integrations that my company offers. It also allows partners and customers to develop and build their own custom integrations and technical solutions for the platform.

Within months of starting my new role and with API documentation high on the agenda, the project was soon assigned to me to get underway.

Being honest about it, I wasn't really sure how to start at all. It was a daunting, intimidating task. I wasn't very familiar with APIs. I certainly wasn't familiar with the company's API. In fact, I was still getting familiar with the company's platform and offering as a whole. I was a true beginner in this respect and it felt scary.

The problems with our API documentation as it was then

The key problems were as follows.

Disparateness

Any available documentation was scattered around and fragmented. Documentation for the different versions was located in different places. Sometimes the documentation for the different versions was located in the *same* places. For instance, our marketing site (but not our support site) had our established help pages on version 1.0, there was an independent page that hosted the auto-generated documentation for version 2.0 (in essence one big list of all of our SOAP methods and REST operations), while our support site contained the odd article here and there on both API versions. The forums on our support site also had discussions and feature requests on both versions 1.0 and 2.0 and it wasn't necessarily clear which. This meant documentation was unnavigable, hard to comprehend and difficult to determine whether it was relevant and up to date.

Unfriendly auto-generated documentation

Our API version 2.0 consisted of auto-generated documentation. This was impenetrable, dynamically generated content, with a pedestrian design that was straight out of XSLT (Extensible Stylesheet Language Transformations) hell. It was ugly and it wasn't user friendly.

It was unreadable to the untrained eye, literally because there was barely anything to read in the way of sentences or descriptions. It served as a sterile, machine-readable list of all methods and operations, featuring endpoints and a brief description of what each method and operation did (see Figure 1).

Summary

Resources	Method	Description
https://api.dotmailer.com/v2/account-info	GET	Gets a summary of information about the current status of the account.
https://api.dotmailer.com/v2/address-books	POST	Creates an address book.
https://api.dotmailer.com/v2/address-books/ {addressBookId}/campaigns?select={select}& skip={skip}	GET	Gets any campaigns that have been sent to an address book.
https://api.dotmailer.com/v2/address-books/ {addressBookId}/contacts	DELETE	Deletes all contacts from a given address book.
	POST	Adds a contact to a given address book.
https://api.dotmailer.com/v2/address-books/ {addressBookId}/contacts/{contactId}	DELETE	Deletes a contact from a given address book.
https://api.dotmailer.com/v2/address-books/ {addressBookId}/contacts/delete	POST	Deletes multiple contacts from an address book. This will run in the background, and for larger address books may take some minutes to fully delete.

Figure 1: The top level of our auto-generated REST API documentation, listing its operations

Clicking the endpoint jumped to the object name in a request and response. The object name then needed to be clicked on to jump to its list of parameters. If parameters had default values, then another link needed to be clicked on to jump to those (see Figure 2). From an information design and user experience perspective, it was a bit of a nightmare.

ApiAddressBook

Field	Type	Description
id	xs:int	
Name	xs:string	
Visibility	ApiAddressBookVisibility	
Contacts	xs:int	

Figure 2: A list of body parameters for the address book object in our auto-generated REST API documentation. Clicking the 'Visibility' parameter type jumped to a list of the parameter's default values.

This was an API reference only (and not a very good one at that), with no overview or general context around it, nor did it look relatable to anything a developer might be doing or seeing within their IDE (integrated development environment). It was written by a machine for a machine. It was clinical, unattractive and bereft of description, voice or tone. As many developers have often said to me, they want API documentation that looks like a human wrote it with the full intention of having another human read it! This wasn't it.

Lack of a documentation owner

It was clear no one was in charge of maintaining API documentation across the board. From what I could gather, our developers had set up an auto-generated WADL (Web Application Description Language) for REST and WSDL (Web Services Description Language) for SOAP to keep API version 2.0 documentation automatically up to date. When I say 'up to date', it was to the extent that if a new operation or method was added to the API, it would then automatically get added to the list with the others. Understandably, this suited our developers (and the company), enabling them to spend their time working on things other than documentation.

In terms of the API version 1.0 documentation, this was deprecated content and nothing further was going to be done with it. Any other activity around API documentation content involved various members of the company, from various departments and with varying experience of the API, contributing the odd support article here and there about a specific issue. Various team members also replied now and then to support forum discussions on the subject of the API, which could be about both version 1.0 or 2.0.

A single owner was needed, one who could marshal and centralize all of this content and would be responsible for the collation and upkeep of it.

Questions I needed to consider – and obstacles I needed to overcome

With increased uptake and usage of version 2.0 of the API, the company had to improve on what it had.

There were some key questions that needed to be answered and obstacles that needed to be overcome:

- What would the new documentation look like in terms of design? What would we opt for in terms of style and presentation?

- What would the documentation consist of with regards to content? As well as the API reference, what else would we include? An overview? Code samples? If so, how detailed? Scenario-based tutorials? A community forum, allowing for a feedback loop?

- How were we going to publish and maintain our new documentation? Writing it was one thing; how were we then going to deliver it?

- Who would be in charge of documentation and its upkeep? What would the process for upkeep be?

- An obstacle from a company viewpoint was devoting adequate resource, focus and support to completing API documentation. I was going to need support, after all. I was going to need help with knowledge transfer and generation of code samples, not to mention the time of designers to design the documentation. How easily could the company be able to justify redirecting the required resource in light of other business demands? Although something of a business priority, would the company be able to avoid relegating API documentation for other pressing priorities?

- An obstacle from my viewpoint was my lack of experience and confidence in this area. I was a writer, not a developer. This was outside of anything I'd done before. How was I going to learn the whole of our API, in both REST and SOAP, to the level of expertise that allowed me to then document it better than anyone had done before? Would I feel overwhelmed by the task and be tempted to relegate it myself for other pressing priorities?

How everything was resolved

Although I'm going to explain how all of these questions and obstacles were resolved in as clear and succinct a way as possible, the reality was a lot messier. It all took place over a fragmented timeline of around two and a bit years, with the project looming and receding at various stages during business-as-usual activities. Decisions were made and elements were worked upon at various times, while other parts of the project encountered delays and dead ends. This in itself serves to underline a more general problem; that of not giving documentation enough focus, nor making it enough of a priority, nor giving it a firm deadline. In the quick-moving environment of a busy development team, documentation can all too easily find itself taking a back seat.

What would the new documentation look like?

I wanted it to have a modern look and feel and to move as far away as possible from our auto-generated documentation. I wanted it to look like it was a part of our support site, with a clear, clean design – friendly to read and easy to navigate. I was thinking along the lines of what ReadMe.io's blog (2016a) defines as a "modern layout", to which they attribute essential features such as a multi-column layout, a syntax highlighter for code samples, a persistent navigation bar, plus tabs for code samples in different programming languages.

Stripe's API documentation is an excellent example of all that a modern layout can be, employing all of these features. I highly recommend having a look.

What would the new documentation include?

I decided the new documentation would include:

- **General API overview material.** This would comprise a 'Getting started with the API' article and an FAQ section to cover areas such as setting up an API user, API restrictions and limitations, secure access and authentication, as well as how to get testing with the API.
- **A detailed API reference.** I wanted to give more than just a brief description. I wanted to provide detail on how parameters were best used, give definitions on values, outline any limits and restrictions that may not be obvious, as well as point out any other quirks. I also decided to add an example request and response in JSON (JavaScript Object Notation) for every REST operation, allowing developers to understand the objects they'd be working with. I wanted to remove as much obscurity and frustration for users of the reference as possible.
- **Code samples for every REST operation and SOAP method.** This would be in just one language to begin with, then added to over time. The code samples would be basic ones, to the level that a junior developer could easily work with them. I felt code samples were important, allowing developers to easily cut and paste and then build upon them.

- **Error response types documentation.** I didn't want to just broadly explain what HTTP response codes meant. I wanted to list all of our error response types and provide explanations for them, giving developers the best possible chance to debug and troubleshoot their code.

- **Interactive documentation.** When looking around online, I had seen how good API documentation was increasingly allowing for operations to be tested within it. No tools were required. This made good sense to me, allowing anyone, not just developers who can code, to play around practically with the API and understand what it can do. After all, it may not be just developers looking at the documentation. It could be members of a QA team, a product manager or even a CEO of another company.

- **A line of communication.** Documentation can be a bit of a one-way conversation. It's frustrating if a developer spots an error, something that could be explained better or something they feel is missing but they don't then have a quick line of communication to the author. I wanted to create a feedback loop. After all, developers using your documentation are the best proofreaders you have.

On this last point, I knew we were going to be in a good position. Our marketing site already had a third-party live chat function available and I knew this would be added to our launched documentation too. I would be the respondent.

Live chat was invaluable, if a little frightening at first. Supplying documentation that you've researched is one thing. Fielding live questions from actual API users is quite another. However, it enabled and accelerated my learning in two distinct areas: I learnt even more about the nuts-and-bolts working of the API (by going off and using it to look into what the user wanted to know), plus I could instantly gauge what our users were struggling to understand. They could also provide feedback on the documentation and speak to someone while using it. I was then able to edit and improve the documentation based on this feedback loop. If I was unable to answer the question, I simply converted the conversation into a support ticket for our dedicated support team to handle. If you're launching API documentation or a developer hub, I would strongly recommend employing this feature.

How was the new documentation going to be delivered?

A number of options were considered:

1 **Use the same third-party platform which we used for our knowledge base.** However, it wasn't really tooled for a satisfying developer experience and I thought it would ultimately prove unwieldy for good API documentation.

2 **Use Swagger.** I considered this when I first began looking around the web for guidance in 2013. Swagger (or the Open API Initiative, as it's now known) is a popular, language-agnostic specification for describing REST APIs. I was inspired by the clean and attractive look of the documentation it could produce, as well as the interactivity it supported. Operations are clearly and neatly listed and can be individually expanded to reveal a full description of their parameters. From there, calls can be made and responses can be seen.

   However, as a non-developer myself, it wasn't particularly accessible or practical for me. It soon became apparent that it would take weeks to learn to use the framework. In short, I didn't want to spend additional time learning a description language in order to document our API. Additionally, we already had an existing API designed and in use, whereas the Swagger framework appears more suitable for an organization looking to design, describe and document their API from scratch. From another standpoint – and Tom Johnson (2016) highlights this in his evaluation of Swagger – the framework may well excel in API reference documentation but that's where it ends. Overview and tutorial content would need to be delivered with an alternative solution.

   That said, I drew a clearer vision of the direction in which I wanted our documentation to take thanks to Swagger.

3 **Produce our own site and our own design, that makes use of XSLT, to publish ourselves.** This option would require more resource than just me. It would also need significant design and development time.

4 **Find and use a third-party platform.** This would ideally provide me with all the tools so I could do the things I wanted to do, without the company needing to design or develop it, nor requiring me to learn a specification or framework.

When first looking in 2013, the fourth option didn't seem to be around, but it was when we next picked up the project. We opted for ReadMe.io – a SaaS site allowing for the easy publication and upkeep of professional-looking content via a platform purposely designed for API and developer documentation. It met all of our needs, plus it allowed the company to switch from thinking about mere API documentation to launching a full developer hub instead.

Who would be in charge of the new documentation and its upkeep?

This was simple. I would be the owner and champion of the documentation. As the company's technical writer, my job title tended to dictate that! However, asking this question is a good thing, I believe. Could you have a number of people writing and updating documentation, including in-the-know developers, in the spirit of 'many hands make light work'? Maybe but I think that would be hard to manage and for the documentation to remain consistent. The best strategy for us was to have a single owner, a single point of contact who ensured documentation standards were maintained. Contributions, updates and suggestions would go through me before publishing them. The development team – and of course anyone else for that matter – could feed into this. The idea was that it would avoid a documentation free-for-all, not unlike the situation the company had previously got into.

Also, as part of an Agile development team working in fortnightly sprints and holding sprint reviews, I would always be aware of any changes to the API before they were deployed. I was perfectly placed to keep documentation up to date.

Overcoming the company's obstacle – providing resource and focus

The following may sound ridiculously simple, maybe because it is and it gets overlooked: my line manager ensured revamped API documentation became an official goal to be reviewed at appraisal time, setting a firm deadline. This in turn aligned with an official overall company goal: enabling the adoption of technology for our users. It suddenly provided the focus and impetus that had previously been lacking. Rather than skulking in the shadows as something of a side project, API documentation became fully recognized by me and the company as a priority. I was duly assigned developer and designer resource and all of my other work was parked so I could concentrate solely on it.

We committed to a launch date that coincided with a major upgrade of the platform, tying in with a revamp of the overall look of our support content. This enabled it to become a big announcement and something to shout about. It became central to the narrative for the launch and our marketing communications – 'More helpful help'.

Overcoming my obstacle – being a beginner

There is no secret or shortcut to this. Faithfully documenting an API requires a knowledge of it and how it works. That has to be acquired by *doing*. From my scouring of the web, it looked like documenting an API was acquired by doing too. In my position, this was a frustrating chicken-and-egg scenario.

The web, as ever, is a trusty go-to resource (just ask a developer when they want to solve a coding problem!). However, when I was first looking around in 2013, there wasn't an awful lot of guidance on the subject. There's a lot more now and there are even online courses on API documentation that can be taken, such as Tom Johnson's (2015a). Those courses could go a long way to solving the chicken-and-egg problem for beginners like me in future.

To begin with though, prior to discovering ReadMe.io, I got my hands dirty by spending hours, days and weeks working through all of our REST operations and SOAP methods, of which there are over 100 of each respectively. Some were easier to test than others. For instance, retrieving a list of data with a GET was straightforward but posting or updating complex transactional data wasn't.

To get to grips with all of these, I used Chrome's Postman (for REST) and SmartBear's SoapUI (for SOAP). They enable interfacing with APIs. You can quickly add a WADL and WSDL respectively to start using all of their operations and methods. I made use of two accounts with which to experiment and test: my account, as well as a dummy account with dummy data in it that our developers had created (and we provide the details of this dummy account in our documentation so users can also play around with it).

I initially documented everything about the operations and methods in an Excel spreadsheet. You may well ask why I recorded it in that format. It was because we thought we might create and publish our own content at the time, with this format making the most sense for the way in which it would get imported into the system.

By the end of this testing process, I knew the API pretty well inside out. I also had a lot of my documentation written, including an overview and FAQs (in a normal Word document this time). The only problem was that the API reference was locked away in an unusable and unreadable spreadsheet.

The discovery of ReadMe.io, however, soon solved this.

Working with ReadMe.io

ReadMe.io was perfect for both mine and the company's needs.

Thanks to its intuitive user interface, I was quickly able to convert the content I had in my spreadsheet into tangible, structured and attractive documentation (see Figure 3 on page 192).

This proved extremely satisfying from a personal perspective, having had all that work previously squirrelled away in a spreadsheet for months on end. Finally, it was being unleashed and I could see what I was working towards.

Using a third-party platform made sense for the company too. Rather than invest hours in creating and designing our own site, the hard work was already done for us, saving us hours in development and design time.

What exactly, then, made ReadMe.io a great choice for me and for delivering the standard of documentation that I wanted to deliver?

Simple, intuitive user interface that's easy to manipulate

As a non-developer and non-designer, this felt like a writer's tool with no barrier to entry, unlike Swagger. I was able to cut and paste content into it that I already had prepared elsewhere, while it features a reassuring Markdown editor. Various elements can also be dragged and dropped into your documentation from a side panel, such as headers, callout boxes, tables, images, embed URLs and code samples.

REST documentation templates (and other choices)

When creating a new document, you can choose from a tutorial, function, FAQ or external URL template. You can also select a REST verb from a dropdown to create a REST template that has everything you need on it

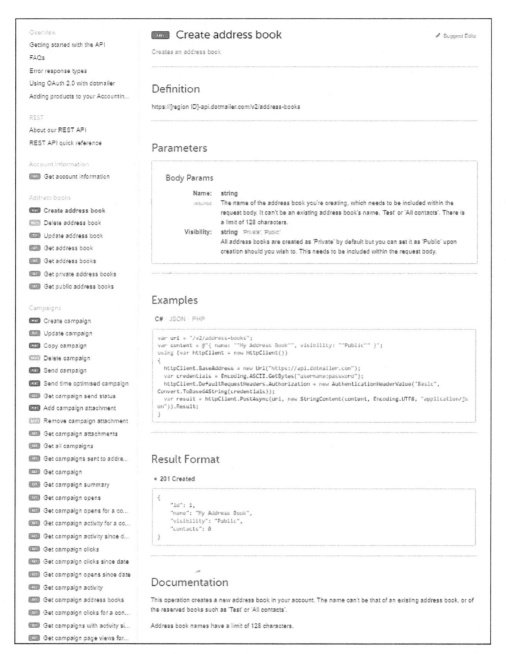

Figure 3: Our REST API's 'Create address book' operation documented on ReadMe.io's platform, featuring full explanations of parameters, some code examples, a response in JSON and any further information that might be useful.

to create great-looking documentation for all of your operations. The template of course enables you to include a name for the operation and a description, followed by the operation's URL. You can add path, query, body and header parameters, depending on your needs, then describe them in detail. You provide the parameter's name, mark whether it's required or not, select its data type from a dropdown list, add any default parameter values, then include a description of the parameter. You can then add an example request and response in just about any form or programming language you want. Underneath this there is room for further documentation of the operation.

Interactive documentation

All of the above feeds into the ReadMe.io's API Explorer functionality. This lets users try out REST operations from within the API reference documentation, provided the user has valid API credentials. They can add values for the operation's parameters and then make the call, which will provide a JSON response. This is a genuine call to the API and the user's account, so if something is being posted, updated or deleted, it will genuinely be executed. This needs to be understood by the user, so I provided a prominent warning to that effect. However, it's a great way to enable readers of the documentation to quickly test an operation out, understand it more fully and get to grips with a response.

Just as a side note, I had to disable the API Explorer a few months after launch. This was because our base URL for endpoints became region-specific depending upon an account's locale. As such, we were no longer able to provide a single base URL in the documentation, in case it caused confusion for users in a different region. Thus the API Explorer was no longer an option for us but it's something I'd ideally love to bring back.

Suggested edits from the community

Anyone can create a ReadMe.io account, which means any reader of the documentation can make use of the 'Suggested Edits' link available on every page. If they see a mistake or if they feel there's information not being covered on the page, they can click on the link to submit their suggestion. A notification is then given to me in my ReadMe.io project's dashboard, where I can choose to act on the suggestion or dismiss it.

This facility is hugely useful and I was – and always am – thankful for these suggestions. Being able to crowdsource from a base of developers makes perfect sense. It can potentially save a lot of time and energy having an army of proofreaders and testers willing and able to feedback to me. It also provides our community of developers with a feeling of investment in the documentation when they see their suggestions go live.

Ability to customize appearance and create a landing page

Our designer could easily customize the appearance of the developer hub, making sure it was company-branded and looked like a deliberate part of our offering. This lack of design and branding is certainly an area in which our previous API documentation – and other companies' – can fall down, coming across as something entirely separate.

The ability to create a landing page is a key part of this customization and branding too. It's a bold statement for visitors to arrive at, declaring itself loudly and proudly – and why shouldn't that be the case? If you've created API documentation, why should it hide away in the shadows, apologetic for its existence? Our landing page clearly looks like our help centre, with a prominent 'Get started' button to take first-time visitors straight to the 'Getting started with the API' material. The rest of the page contains prominent links to all of the other pages of the developer hub, including the API operations and methods, making the navigation obvious immediately (see Figure 4).

Figure 4: Our developer hub's customized landing page on ReadMe.io's platform

All in all, ReadMe.io is an extremely user-friendly and flexible documentation tool, allowing for swift generation of documentation. It provided a solution to many of our problems and I can continue to update, extend and maintain it to my heart's content.

I also needed to document SOAP methods too. Despite there being no SOAP template, I was still able to do this effectively using ReadMe.io's tools.

Projections for the future

Writing and researching this chapter has been yet another learning process. It's renewed my desire to return to the documentation and build upon it further: I'm the 'expert' who *was* a beginner but who realises he's *still* only *really* a beginner and wants to become *that* expert. To a degree, that's probably where overall thinking on API documentation is right now, especially as it's still bereft of a standard.

There's a gathering drive towards a documentation standard, though. Recent years have seen increased research going into API documentation and it looks set to continue, in line with the desire for greater understanding of the needs of those using it.

The fact that API documentation deserves this attention can't be ignored. A survey by Programmable Web (2013) of 250 respondents (largely developers) found that complete and accurate documentation was the most important factor in an API. This came above service availability and uptime.

Drilling down on documentation shortcomings, Gias Uddin and Martin P. Robillard's study (2015) 'How API Documentation Fails' surveyed over 300 IBM software professionals to find out that their top-rated problem was incompleteness of API documentation, followed by incorrectness and then ambiguity respectively. Their major finding, however, was that respondents cared most about quality content – and they flagged up how this is the hardest aspect to solve. Complete and correct documentation requires an expert user of the API, which is likely to be a developer. A non-developer will find acquiring this information difficult. Developers often don't have the time to devote to this. One solution Uddin and Robillard suggest is to harness the power of the developer community to quickly identify what needs to be updated. The developer community is clearly set to play a major role in shaping the development of any API's documentation.

Michael Meng's study 'How API Documentation Can Be Effective and Useful' (Mantke, 2016 and Write the Docs, 2016) uncovered similar findings to those of Uddin and Robillard. Meng surveyed developers by asking questions about what they looked for in API documentation, how they used it and the problems they found with it. He discovered that users were looking for complete and up-to-date documentation, with detailed explanations of operations and code samples to go with them. Typical problems were that documentation was wrong, incomplete or incomprehensible.

Meng made two particularly telling findings, however, both of which would appear to point to how API documentation is set to develop and mature.

Firstly, he found there were two distinct requirements from the users of the documentation: the need for conceptual guidance (a general overview of the purpose of the API and what can be done with it) and practical guidance (the specifics of individual operations and parameters and how they can be used). In his opinion, API documentation needs to cater for both requirements and needs to be flexible in offering support for different API learning strategies.

Secondly, he used eye-tracking software to observe how developers made use of API documentation when charged with completing a task. This revealed that they weren't reading content through but were scanning quickly for answers to their problems (like most readers of online information), such as looking for headings and keywords. It highlights how important good information design is, with clear, consistent navigation that enables speedy problem solving. For example, Meng proposed that API documentation would be improved with content-based classification and titles (for example, 'Creating contacts and address books') as opposed to more general document types ('Overview', 'Tutorials', etc.).

We're moving away from the days of development being a 'closed shop'. More and more people are turning their hand to development, even in their spare time. The development of systems in a microservices architecture is also becoming hugely popular. This proliferation and integration of a lot of smaller services will require more documentation, which in turn must become more sophisticated and mature to support this need effectively.

As APIs evolve and mature, API documentation will do the same, meaning the writers of API documentation will need to evolve too. The skillset required is likely to broaden significantly to encompass the convergence of technical writing, information design and user experience.

APIs are also increasingly productising and with this productization come two things:

- API documentation becomes the interface through which users engage with the API product
- API documentation also becomes the pre-sales and marketing content that a prospective customer uses to decide whether the product is for them.

This means writers will need to balance detailed technical information with broader, more conceptual information – as Meng's study identified – and become skilled in writing in a lighter language and tone that doesn't scare off a potential purchaser of the services.

Conclusion

Writing good API documentation is a challenge for various reasons, none more so than when starting out as a beginner.

However, planning is the key, as well as making use of the resources that are out there, whether it be online advice from API documentation thought leaders (and these are well worth subscribing to and following for the latest developments in the field) or using a SaaS documentation platform such as I used.

It's important not to be daunted or intimidated by the task, nor by the various obstacles that might be encountered, as have been discussed. It would be wonderful to anticipate a developer's every need and launch perfect documentation that caters accordingly. However, it's far from realistic. The important thing is to firstly get up and running, something that ReadMe.io (2016) identify as "minimum viable documentation", on a platform that is ideally future-proofed: something that can be easily edited, extended and built upon. One of the biggest learning processes comes *after* the documentation is live: the feedback from users. Any missing pieces can be provided once documentation is out in the wild, backed up by regular reviews of it.

At the time of writing, it's been just over a year since the launch of our developer hub. There have been numerous compliments to its look and feel, coverage, clarity and the availability of code samples. How have I been able to gauge the external reaction? Largely thanks to feedback via the live chat facility – and those compliments are, of course, gratifying. One thing is certain; our developer hub is light years better than our previous API documentation offering – and that's a great asset for the company to have.

Am I satisfied? By no means. The documentation isn't perfect; there's room for improvement and extension. I'd like to provide code samples in more programming languages (we've already added PHP), scenario-based tutorials, possibly some quickstart guides – the list goes on.

I'd ideally like to aspire towards Ori Pekelman's 3:30:3 rule. This rule is the measure of whether an API is easy to set up and use, describing the best possible engagement between the documentation and the reader. The holy grail is that a landing page manages to convey what an API does in three seconds; it takes no more than 30 seconds to find the correct endpoint required; then an account can be created, a call made and the result received in under three minutes.

Improvements aside though, our developer hub was still successfully launched and there's now a very good standard established for building upon. And that's what API documentation should be – an iterative process as you learn more and more about your users' needs; living and breathing documentation.

While there is no standard for API documentation yet, nor a golden handbook or sure-fire formula for producing good API documentation, this discussion has covered fundamental cornerstones and emerging best practices associated with it.

- **Attractive**, clean and branded presentation, so it can be easily read and looks like an intentional part of your documentation offering
- **Easily navigable**, with a two- or three-pane layout that allows for as much information on the screen as possible
- **A conceptual overview**, featuring what the API can be used for and what can be achieved with it, while covering authentication, restrictions, testing of the API and other resources
- **An FAQ**, covering the troubleshooting of common issues
- **A detailed API reference**, including full documentation of all methods and operations, covering all parameters, requests and responses, plus any limitations and quirks

- **Interactive documentation**, providing the ability to try calls right there within the documentation

- **Code samples per method or operation** and ideally in as many languages as you can supply

- **A full list of all error response types and explanations of them**, enabling quicker debugging and troubleshooting of code

- **Tutorials covering common use cases and scenarios**, adding greater depth to the documentation and joining up dots so developers can get started quicker

- **Harness the community**, by ideally providing a community forum or live chat or any sort of feedback loop so it can be channelled back into the documentation.

Lastly, it will certainly help if a technical writer pushes for all of the above, takes ownership and champions the API documentation project as a whole. It's likely they – and the documentation – will become some of the greatest assets the company has.

How did I arrive at this 'wisdom'? Purely from having done it the once. There is no other way to learn. To quote Aristotle; "For the things we have to learn before we can do them, we learn by doing them." There can be great consolation in philosophy and I have to say I needed to turn to it a fair few times as I documented our API!

As documenting an API is a continual learning process, in my opinion, it's also worth keeping in mind Socrates' humbling maxim; "The only true wisdom is in knowing you know nothing." This should keep any technical writer returning regularly to review their documentation, to ensure it's as good as it possibly can be.

References

Biehl, M. (2016). *API Design: APIs your consumers will love*. Great Britain: Amazon.

DuVander, A. (2013). *API consumers want reliability, documentation and community*, ProgrammableWeb. http://www.programmableweb.com/news/api-consumers-want-reliability-documentation-and-community/2013/01/07 [Accessed September 2015].

Johnson, T. (2016). Implementing Swagger with API docs. *Communicator*, Autumn 2016, pp. 43–47.

Johnson, T. (2015a). *Documenting REST APIs*, http://idratherbewriting.com/docapis_course_overview [Accessed October 2015].

Johnson, T. (2015b). *Most important factor in APIs is complete and accurate documentation.* http://idratherbewriting.com/2015/01/15/most-important-factor-in-apis-is-complete-and-accurate-documentation [Accessed October 2015].

Jensen, C. T. (2015). *APIs for Dummies: IBM Limited Edition.* New Jersey: John Wiley & Sons, Inc.

Lane, K. (2013). *History of APIs*, API Evangelist. http://history.apievangelist.com/ [Accessed November 2016].

Mantke, C. (2016). *Poll results: What should good API documentation do?* (in German), JUG Saxony. https://jugsaxony.org/2016/07/21/ergebnisse-der-umfrage-was-muss-gute-api-dokumentation-leisten [Accessed November 2016].

Nijim, S. and Pagano, B. (2014). *APIs for Dummies: Apigee Special Edition.* New Jersey: John Wiley & Sons, Inc.

Pedro, B. (2014). *5 reasons why developers are not using your API*, Nordic APIs. http://nordicapis.com/5-reasons-why-developers-are-not-using-your-api/ [Accessed November 2016].

Postman: https://www.getpostman.com.

Pratt, E. (2016). *Research into how API documentation fails*, Cherryleaf blog. https://www.cherryleaf.com/blog/2016/09/research-into-how-api-documentation-fails [Accessed November 2016].

ReadMe.io: http://readme.io [Accessed July 2015].

ReadMe.io blog: https://blog.readme.io [Accessed November 2016].

ReadMe.io blog (2016). *Best practices for writing API docs and keeping them up to date*, 19 September 2016. https://blog.readme.io/best-practices-for-writing-api-docs-and-keeping-them-up-to-date [Accessed November 2016].

Robillard, M. P. and Uddin, G. (2015). How API documentation fails. *IEEE Software*, 32(4). Available at: http://www.cs.mcgill.ca/~martin/papers/ieeesw2015.pdf [Accessed November 2016].

SoapUI: https://www.soapui.org.

Stripe API documentation: https://stripe.com/docs/api [Accessed July 2015].

Sturgeon, P. (2015). *Build APIs You Won't Hate: Everyone and their dog wants an API, so you should probably learn how to build one.* Great Britain: Amazon.

Swagger petstore demo: http://petstore.swagger.io [Accessed May 2013].

Twilio API documentation: https://www.twilio.com/docs/api [Accessed November 2016].

Write the Docs (2016). *Michael Meng - API documentation: Exploring the information needs of software developers.* Available at: https://www.youtube.com/watch?v=soQSOBwiXdA [Accessed November 2016].

10

The development of DITA XML and the need for effective content reuse

Keith Schengili-Roberts

Abstract   DITA XML is the fastest-growing standard for structured content. In the decade since its launch, DITA has changed the way many organizations undertake their own technical documentation, as well as changed the discourse around how structured content can better communicate consistent messaging in an efficient and cost-effective manner.

Keywords   DITA XML; DITA usage; history of DITA; DITA growth; structured content

Introduction

Currently in use by at over 650 firms worldwide (DITAWriter, 2016), this chapter will examine the developmental roots of DITA and its start at IBM, and how some of the original architectural decisions that were made have opened a dialog about the nature of effective technical communications. This chapter will also explore the leading reasons as to how, where and why DITA was initially adopted, focusing on its advantages from the perspective of technical writers and businesses. DITA has effectively been twinned with business principles such as Agile, and makes possible other business process efficiencies such as lowered localization costs. The chapter will compare some of the key differences between DITA and other structured content formats, and delineate what makes a specification successful from a market and client perspective. It will also describe some of the fundamentals of DITA and provide a perspective on where it stands within the realm of technical communications in general. It will conclude by investigating current trends in the related software tools industry and project where DITA is going with the ongoing development of both DITA 2.0 and Lightweight DITA.

Understanding DITA

The Darwin Information Typing Architecture (DITA) is a XML standard that has been widely adopted within the technical writing community, and has been acknowledged as the fastest-growing technical documentation standard. It is an open standard that was originally released by the Organization for the Advancement of Structured Information Standards (OASIS) back in June 2005, and since then it has been adopted by technical writing departments in at least 650 companies world-wide. It is topic-based and focuses on content reuse, two factors that have helped establish its rapid adoption, as it makes for more consistent messaging, efficient writing processes, and cheaper localization costs. It has seen the highest adoption rates within the software and technology sectors, and is spreading to other sectors as technical writers take their experience to new jobs at other companies. The granular nature of DITA topics works well within firms that have adopted Agile-based processes, and when twinned with a Component Content Management System (CCMS), DITA also lends itself well to production measurement, so technical writing managers can gain a clear sense as to the progress of a documentation project. DITA continues to change and evolve. The latest version of the standard at the time of writing (February 2017) is DITA 1.3, with ongoing work within OASIS to create the next full version of the standard (currently targeted as 'DITA 2.0'), and Lightweight DITA, which seeks to produce a slimmed-down version of the DITA 1.3 standard that can be written in forms other than XML, such as in HTML and Markdown.

The acronym DITA stands for its chief components:

- 'D' is for 'Darwin', alluding to Charles Darwin's concept of evolutionary adaptation, as the structure of DITA is based on the ideas of specialization and inheritance
- 'IT' is for 'Information Typing', defining the semantic structure of individual topics
- 'A' is for 'Architecture', anchoring it in the idea that this is a structured standard that is also an extensible.

The ideas of specialization and inheritance are central to the standard, as all DITA topic types and elements exist in a parent/child relationship. All of the five main topic types are derived ultimately from a generic 'topic' topic type, and all of them share the same basic title and optional body content structure from it. The individual topic types – concept, task, reference, glossary and troubleshooting – are all specialized from the generic 'topic' topic type. They all share an XML identifier, a title and optional body content structure from topic, and then are specialized to suit the types of content that can be placed within them. Similarly, all DITA elements share (or 'inherit') the properties of their parent. Many elements that are commonly used across the various specialized topic types are derived from the 'root' generic topic type. The paragraph (p) element is present in the generic topic type, and so it is inherited by the other topic types derived from it. The more specialized the element, the deeper it is within the hierarchy, so the trouble solution (troubleSolution) element, which is a type of section found only in the troubleshooting topic type, is derived from topic/section troubleshooting/.

Information typing serves as a form of focus for the writer, and provides a structure to convey content to the user. The intent of a concept topic is to provide information on the function of a product features. A task topic provides step-wise instructions on how to do something. A reference topic provides additional supporting information that may be of use or interest, such as the range or optimal settings for a product feature. The troubleshooting topic provides a solution to a specific issue a user may encounter while operating the product, and a glossary topic provides a description of terms pertaining to the product. All of these topics are referenced using a map, which contains links to all of the topics, as well as optional metadata and other related content comprising a document.

Another design philosophy incorporated into DITA is content reuse. The topic-based nature of DITA means that it is modular, so it is possible to use the same topics in different maps. When used properly, this approach ensures consistent messaging. The modular nature of topics and the way they can be swapped in and out of maps means that writers have to take a non-narrative approach when writing content for a topic, as they cannot be sure of the placement of a given topic within a map. As a result, a best practice is to ensure DITA topics can be read by users as standalone content, concisely conveying exactly what a user needs to know for a given scenario. Content reuse also occurs below the topic level, with conrefs ('content references') and keys which can be used interchangeably for handling content reuse at the paragraph, sentence, phrase or individual word level. This ability to reuse content at a very granular level as well as at the topic level is one of the most distinguishing features of the DITA standard, setting it apart from other structured documentation standards.

Content reuse also greatly reduces the costs associated with translating content. If, for example, English source topics that have already been translated are used in a subsequent publication and Translation Memory (software used for keeping track of previous translation work) is used, the amount of new content that needs to be translated is greatly reduced. In some cases, the number of words that need to be translated can be zero. This feature of DITA makes it attractive for companies seeking an efficient and cost-effective way to produce content in multiple languages.

Several formatting output types (including XHTML, PDF, HTML5, CHM and others) are available to the writer when they want to 'publish' their DITA content. DITA XML separates form from content, so the XML content is transformed using the Extensible Stylesheet Language (XSL) to produce content in the desired target format. The DITA Open Toolkit (DITA-OT) is a freely available, open-source publication engine that is most commonly used to produce content in the desired target format. The separation of form from content provides additional efficiency to the technical writing staff, as they no longer have to spend time formatting content – such as changing header levels, the fonts being used, tweaking table spacing or other related operations – as the output engine takes care of that automatically as the document is outputted to the chosen target format. This applies also to translated content, so long as multilingual fonts are available to the DITA-OT at output. Professional formatting tools are available to aid with the more complex operations relating to translated content, such as language-specific hyphenation rules, punctuation, kerning, ligatures, working with right-to-left languages and more.

The origin and spread of DITA

The origin of DITA goes back more than a decade with IBM's pioneering efforts in topic-based authoring. IBM started to create structured content using their own Information Structure Identification Language (ISIL) even before the advent of the Standard Generalized Markup Language (SGML) (DITAWriter, 2016). The advent of the then-new XML standard that was released back in 1998 (Wikipedia, n.d.) provided an opportunity for the technical writing teams within IBM to do a fresh assessment of their needs, especially given the ascendancy of content being published to the

World Wide Web. A lot of the design concepts that went into DITA were derived from IBM's internal guide for developing technical content (*Developing Quality Technical Information: A Handbook for Writers and Editors*), which emphasized minimalism, the separation of conceptual content from tasks, organizing reference content, and how to 'chunk' content into its constituent parts (DITAWriter, 2016). The results of an internal report within IBM recommended that the best way to make use of the greater structural flexibility and simplicity XML offered over SGML, was to adopt a topic-based architecture in order to better support the page-based nature of the web (DITAWriter, 2016). Realizing that they had more to gain by making this an open standard, which would allow IBM to share content to and from their business partners, the ideas behind DITA were circulated publicly in late 2001 with the publication of the white paper *DITA XML: A Reuse by Reference Architecture for Technical Documentation* (Priestley, 2001) authored by Michael Priestley of IBM. An early draft version of DITA was given to the OASIS standards body, leading to the release of the first published DITA standard (version 1.0) in June 2005.

DITA then spread to numerous early adopters, including Research in Motion/BlackBerry and Nokia in the telecommunications sector, AMD in the semiconductor industry, and Siemens Medical in the medical device sector. Other early adopters included Kone (elevator manufacturer), Adobe and IXIASOFT (software), and Xylem (water technology). While the promise of content reuse appealed to a variety of firms, only high-tech companies initially had the tools at hand – such as version control software for primitive, topic-based component content management – to take full advantage of it. The early DITA-adopting firms acted like incubators, with the members of their technical writing teams spreading to an ever-growing number of other firms over time. Best practices in technical documentation processes learned at one firm tend to be carried over to the next, a significant factor in the spread of DITA over time. Based on data from LinkedIn, former employees who worked with DITA at Nokia have gone on to actively using their DITA knowledge at over 30 subsequent companies.

From my research of DITA-using firms, the computer software and the related information technology/services sector comprise 39% of all firms using DITA. This is not surprising, considering that DITA first gained its foothold in these sectors. This is followed by significant numbers of firms in the telecommunication, semiconductor, medical device and financial service sectors. Over the years, DITA adoption within other industries has spread, with significant numbers of firms using DITA in the electrical/electronic manufacturing, mechanical/industrial engineering, aviation/aerospace, automotive machinery and education/e-learning sectors among others.

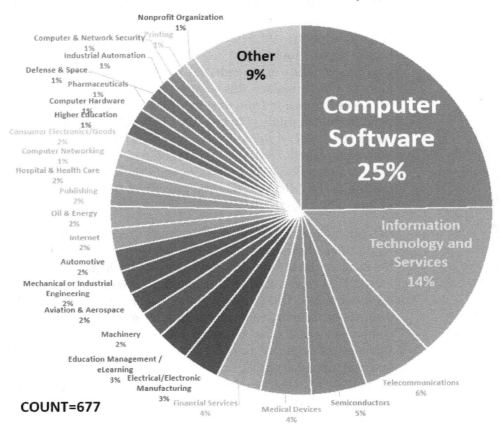

DITA USAGE BY INDUSTRY SECTOR, Q4 2016

COUNT=677

Figure 1: Industry spread of DITA-using firms based on data from DITAWriter

Two-thirds of DITA-using firms are headquartered in North America, with much of the rest coming from Europe (27%) and Asia (7%). There is relatively little adoption of DITA (that is known) in South America, Australia or Africa.

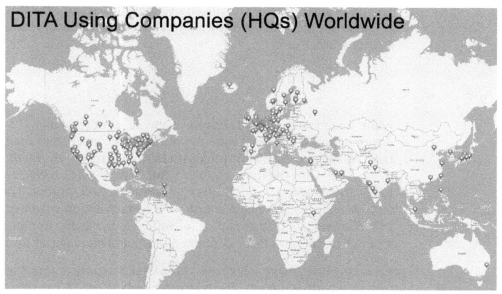

Figure 2: Distribution of the HQs of DITA-using firms worldwide based on data from DITAWriter

Toolsets related to DITA

For most documentation groups, a move to DITA means a change in the way they create, store and publish content. For many documentation teams prior to moving to DITA, traditional word processors (such as Microsoft Word) or desktop publishing programs (such as Adobe FrameMaker) are the most common tools that are used for writing content. This is in addition to supporting tools like TechSmith SnagIt for grabbing screenshots or Adobe Illustrator for creating line drawings. All of these tools can still be used in a DITA environment, but the expectation in most documentation groups is that they will begin to author content directly in XML, store that content within a system designed for optimizing content reuse, and be able to produce content directly for consumption by customers.

There are three primary sets of tools that are typical to most DITA-based documentation teams:

1 An XML editor

2 A Component Content Management System (CCMS)

3 A publication engine for producing output.

The following sections provide an overview describing what these tools are and how they are used within a DITA-based technical writing environment.

XML editors

While there are many XML editors available, a professional documentation team will want a tool that has convenience features tailored specifically for working with DITA. There are several of these on the market, with the most popular being Syncro Soft's oXygen editor, JustSystem's XMetaL and Adobe's FrameMaker XML Author. They all include features that make it easier for technical writers to work with DITA content, such as providing in-context XML tagging options, making it easier to find and reuse content, and to publish material for review.

Taken together these tools have taken over a significant part of the technical writing market. The relative popularity of XML within technical communications signals has been a significant change within the technical writing community. For much of the past quarter century, experience with Adobe FrameMaker was considered to be a primary skill. My research of job postings on the US job aggregator website Indeed.com over the past few years has shown that firms are now seeking technical writers with XML experience over that of FrameMaker, and at a rate that is growing over time. Of the various XML standards used for documentation purposes, technical writer job postings seeking DITA experience comprise the bulk of what firms are seeking in prospective applicants.

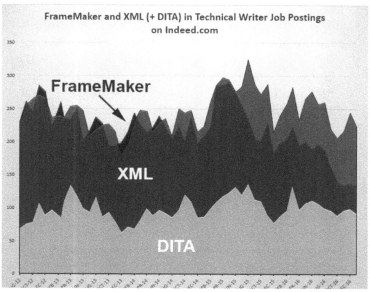

Figure 3: Frequency with which FrameMaker, XML and DITA turn up in technical writer job postings on Indeed.com

Component content management systems

It is rare these days for any documentation team not to use some form of database or repository for storing their technical content. Depending on the number of products a company produces, a documentation team can be expected to generate significant amounts of content over time. If the company localizes its content and there is any expectation of content reuse, standard file-storing options such as shared network folders are soon outgrown. What I have observed in documentation groups over the years is that there often comes a point where any further growth depends on moving away from a simple file folder structure or versioned content repository.

Some small to medium-sized technical documentation groups are moving to content-versioning systems such as GitHub, which in addition to being a repository for content also offers a measure of collaboration. But as this type of system is not optimized for DITA, it does not include features designed to make a technical writer's life easier when using DITA, such as the ability to manage links or find content for reuse. For documentation teams looking for a scalable solution while also implementing DITA, a component content-management system (CCMS) is the answer.

DITA is a topic-based specification, so a CCMS takes the individual documentation components – in other words, topics – and uses DITA maps to provide a hierarchical structure for the content. It then assembles them into documentation deliverables in whatever format is required, such as a PDFs, HTML, eBook and so on. Some of the main DITA CCMSs on the market include SDL Knowledge Center, IXIASOFT DITA CMS, Vasont DITA CMS, easyDITA and DITAToo among others.

A typical DITA CCMS will enable information architects and technical writers to search for individual topics and their contents and then assemble them within a map using a graphical user interface.

Content versioning is another feature common to most content management systems. Originally created for software development, versioning allows users to 'check out' content for authoring and 'check in' any changes. This type of practice was originally designed for programmers, as it allows them to revert to a previous version of their code to help root out bugs. Similarly, a typical CCMS will have versioning capabilities that ensure that only a single technical writer can work on a given topic at any one time, and will also register who made which edit and allow for a quick comparison between versions. Just like in a programming environment, any documentation 'bugs' that are discovered in a later version of a topic can be reverted to an earlier version without the problem.

Another feature that mature CCMSs provide is workflow. In addition to being able to check in/check out versioned content, authors can also notify the system when they consider their work on a topic to be complete. Depending on how the workflow is setup, the topic can then be routed automatically for someone else to review and approve, such as an editor or product manager. Automating workflow introduces many ways to improve documentation processes and quality. In an engineering environment, subject matter experts (SMEs) may be assigned within the system to write content which is then 'polished' by a technical writer, and then sent to another SME for approval. This type of workflow implies that roles can also be assigned to contributors, helping define who can do what types of actions to particular content. CCMS workflows can ensure that all topics within a document have been verified prior to publication, ensuring that internal quality, external requirements – like those belonging to a medical or legal standard – are met. In addition to being able to route content for review, a robust CCMS will have other workflow processes that can be triggered, such as automatically routing the finalized and approved version for localization purposes.

Topic-based authoring allows for much more finely grained measurement of content production. This is critical to know when planning ahead for future documentation releases. A good DITA CCMS will enable a documentation manager to retrieve a wide range of metrics, allowing them to not only effectively measure the return on investment (ROI) for purchasing a CCMS, but also to understand ongoing production and quality issues. As an example of the latter, it should be possible to measure the topic reuse rate, providing the documentation manager with information on whether existing content is being properly leveraged in the production of new content.

Localization savings is often a key driver in the move to DITA and a CCMS. Content reuse in the source language equates roughly to localization reuse, ensuring significant cost savings over time. A mature DITA CCMS will typically include a convenience feature to aid with the 'packaging' of content to be localized and also be able to easily import the translated content back into the system to enable multi-language publishing from within the CCMS.

A typical commercial DITA CCMS will include these features and more. While it is possible to use DITA within a small team using a file folder filled with topics, documentation teams begin to realize significant production efficiencies and the attendant cost-savings when they use a CCMS that has been optimized for use with DITA.

Publication engines and related software for producing output

The third major component of a DITA-based documentation toolchain is the publication engine software. This may consist of one or more software packages that work with each other to deliver the desired publication output type.

The most commonly used publication engine software package is the DITA Open Toolkit (often abbreviated to 'DITA-OT'). It is freely available, open-source software which is designed to process DITA-based content and interpret DITA features according to rules laid out in the official DITA standard. While it is not the only implementation, as it is freely available it is often integrated with many of the most popular XML editors, CCMSs and dynamic publication engines that are available. It is based on Java, which means that it works on multiple platforms, including Windows, Macintosh and Linux. The DITA-OT is able to publish

DITA content to a wide variety of formats, including HTML5, HTML Help, PDF, XHTML, troff and more (Anderson, n.d.). It uses Extensible Stylesheet Language Formatting Objects ('XSL-FO') to transform DITA content to PDF format, and uses cascading style sheets (CSS) to produce the HTML variant output types (Anderson, n.d.).

The DITA-OT was spun off from IBM's software tools division at the same time that the DITA standard was officially released to provide publication software for those wanting to use the new standard. While IBM still donates resources to the ongoing development of the DITA-OT, as an open-source project it receives input from other outside sources as well.

In addition to the DITA-OT, there are additional typesetting software such as AntennaHouse and RenderX, which augment the functionality and provide more precise control over the formatting of published content. These tools also provide additional capabilities for outputted content, providing more robust capabilities for creating optimized PDF output, working with XML-based graphics (SVG) and mathematical equations (MathML), support for the Pantone colour palette and more. They can also be optimized for handling multilingual content, such as being able to properly layout right-to-left languages like Hebrew and Arabic, implementing language-specific hyphenation rules, and the ability to work with multiple languages within the same document. These types of features are required by documentation teams whose documents are published in multiple languages and require greater control over the look-and-feel of PDFs and HTML-based content produced from DITA-based material.

There are also dynamic publishing solutions that are aimed specifically at producing optimized HTML5 content. This includes software products like Congility DITAweb, Zoomin Docs, Fluid Topics and more. A typical feature of these solutions is the ability to transform DITA content to HTML in a responsive layout format, so that the content can easily be read on portable devices such as tablets and smartphones in addition to desktop displays. Some of these publishing solutions also offer tailored search engine functionality, tying in metadata derived from the DITA source and displaying an interface that includes faceted search, enabling users to better refine their search result in order to find the exact content they are looking for. Other possible features may include enhanced tracking of the web content users are reading, the ability to incorporate feedback from users which is delivered directly to the technical writing team, or simply provide a secured environment for publishing content offered as a service.

Many options are available to a documentation team seeking to publish DITA-based content. While publication engines like the DITA-OT are free, there are often associated expenses involved with the setup and configuration of the output types. Documentation teams looking to customize the look-and-feel of their PDF output may have to rely upon a dedicated internal resource or an external consultant to craft the XSL transforms to produce content that conforms with corporate branding. Work on the XSL transforms or CSS is typically applied to a distinct publication set, so that things like user manuals, installation guides, quick start instructions and so on are distinct and easily identifiable as such by the end user. It also is possible to publish DITA content to other formats, such as ePub, Kindle, OpenOffice and more through additional professional publication tools. In the end DITA content can be delivered in most publishing formats that are available.

DITA and other structured XML formats

DITA is not the only option available when it comes to technical writing: DocBook, S1000D and XML formats developed in-house within a company are other viable options. Each of these have their particular niche in the market, but DITA is making an impact on all of them and it has become the most popular of the XML formats used by technical documentation groups.

A long-standing direct competitor to DITA is the DocBook format. Like DITA it is an open specification sanctioned by OASIS, with version 1.0 issued publicly back in November 2002 (OASIS, 2002), though it was originally developed by HAL Computer Systems and the publisher O'Reilly & Associates back in 1991 (Walsh, 2016). From there it was developed further by several computer-related companies, including Novell, DEC, Hewlett Packard and Sun Microsystems for use in creating documentation (York, 2001). It also strives to provide a single-source format for publishing content to multiple formats, including HTML, HTML Help, Unix man pages and PDF. Like DITA, DocBook is a highly structured way to write content, but unlike DITA its focus is at the book-level or article-level rather than on individual topics. This means that content is written as a narrative rather than as discrete, standalone, reusable topics designed to be used anywhere within a given document. DocBook is also designed more for static, unchanging, monolithic content and not around the concept of content reuse which is central to DITA. DocBook superficially shares some structural elements that are similar to DITA at the block level, but has no equivalents to mechanisms such as content references (conrefs) or keys – two of the fundamental reuse mechanisms used within DITA. Several of the firms who contributed to the original development effort for DocBook are now using DITA, including Hewlett Packard, Oracle (which bought Sun Microsystems) and Micro Focus (which acquired Novell) (DITAWriter, 2016). The DocBook specification continues to be developed, with the most recent version ratified in late November 2016. The latest version acknowledges the influence of DITA by including the concept of 'assemblies', which are topic-like constructs that can be used within DocBook (OASIS, 2015), but otherwise has no content reuse mechanisms aimed at a more granular level.

The S1000D technical documentation specification was originally developed within the aerospace sector over 20 years ago, and it is still widely used within that sector, as well as in the defence, ship industry and construction sectors (Aerospace and Defence Industries Association of Europe, n.d.). Like DocBook and DITA, S1000D is an open standard, in this case governed by the Technical Publications Specification Management Group. Unlike DocBook, S1000D does include a mechanism for the reuse of content, known as data modules. These data modules can contain text and/or graphic content, and can be 'plugged in' where needed within any S1000D document (CDG, 2016). There are a number of data module types, roughly analogous to the DITA topic types, and include information that is specific for creating checklists, service bulletins, front matter, parts data, wiring data, learning modules, procedures, faults, information for the crew/operator and more (CDG, 2016). As you can see from this short list, many of the data modules were originally tailored for specific purposes within the aerospace sector which would not apply in more general circumstances. Each data module comes with a unique identifier, called the Data Module Code, which is designed in part as a mechanism for ensuring that the same module does not appear more than once within a single document (CDG, 2016). This points to one of the key differences between DITA and S1000D, which is the granularity of the level of reuse. While S1000D encourages reuse at the data module level (roughly equivalent to a topic within DITA), it does not have mechanisms for intra-data module reuse. The specificity of some of its module types to the aerospace and related industries limits the appeal for its adoption outside of these sectors. Many aerospace firms are now using DITA along with S1000D, though for different documentation sets. There has been at least one concerted attempt to incorporate DITA content within S1000D data modules (LinkedIn, n.d.), but the proposal was not accepted by the Technical Publications Specification Management Group.

There are also companies that have created their own proprietary XML formats for creating technical documentation. Information on these proprietary formats is hard to find, but it appears that many of these were started prior to the advent of DITA, and some share common roots with SGML. It should be remembered that what would one day become DITA originated as an internal documentation standard devised for use within IBM. Unlike proprietary documentation standards, DITA is an open standard, which has led to the development of supporting tools from both commercial developers and the open-source community. As a result, there is broad tool support for DITA, whereas proprietary XML formats often require a significant and continual investment in internal tool development to support. It also means that as new publication formats become common (such as HTML5, ePub, etc.) internal development needs to produce output formats to match. While going with proprietary XML format may have made sense at the time, a company will need to assess whether ongoing development efforts to support it outweigh the cost and benefits of adopting an open documentation standard that comes with commercially available tools.

One of the chief differentiators of DITA when compared to the other documentation standards available is the ability to reuse content at both granular (word, phrase, sentence, topic) and topic/chapter levels. From a practical perspective, it is these multiple stages of reuse that come into play into making DITA a popular standard, making possible the additional advantages of consistent messaging, lower localization costs, and greater efficiencies as writers reuse existing content instead of having to recreate it.

In a relatively short time, DITA has become the most popular XML format for creating technical documentation. In a multi-year survey of technical writer jobs posted to Indeed.com, technical writer positions that asked for DITA experience have outpaced those seeking experience with the other XML documentation formats.

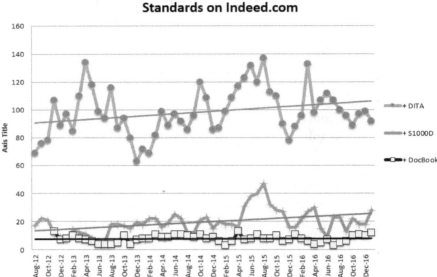

Figure 4: Technical writer job postings and specific XML standards referenced on Indeed.com

Technical writer job postings in the United States seeking those with experience with DITA far outstrips those for the other two XML-based documentation formats. And while technical writer job postings that seek S1000D experience continues to grow, at any given time over the past four years there are roughly 3.5–4 times the number of equivalent postings looking for DITA experience. Technical writer positions where DocBook experience is sought is essentially flat, with some months having the number of such job listings across the US equal to that which you could count on one hand. My advice to any aspiring technical writer these days looking to work with structured content is to learn DITA over the other competing formats.

DITA and Agile

Both Darwin Information Typing Architecture (DITA) and Agile were born out of necessity for software development teams. Technical writers within the software division at IBM established DITA in order to efficiently create effective and collaboratively written topic-based documentation. Similarly, the Agile Manifesto came from a group of software developers seeking a more lightweight way to create their deliverables. Though originally created for different reasons, DITA and

Agile share common roots in the software development world, and DITA can make Agile for documentation teams possible.

Back in the winter of 2001, 17 software developers got together in a remote ski resort to talk about the common issues that they faced with traditional development practices. They published the Agile Manifesto which focused on "uncovering better ways of developing software by doing it and helping others do it" (Beck, 2001). All of the various 'flavours' of Agile – Lean, Scrum, Kanban, Extreme Programming and others – are all ultimately derived from the common principles laid out in the original manifesto, each providing their own take on how to create software more efficiently than waterfall.

Agile software development makes specific demands on documentation teams, whose technical writers are required to be nimble, describe features in a piece-meal fashion, and report on their progress. The structure of DITA is ideally suited to these needs. Agile thrives in environments where short release cycles are possible, and the interviews I have done with members of technical documentation teams who are working within Agile confirms this. Most of those I talked to worked either at a software firm or within the software division of a company in a different industry sector. For example, I have also seen instances of Agile and DITA used together in the medical devices and heavy machinery sectors, but in both cases the push for Agile came from the software divisions within these firms. Agile appears to be rare in highly regulated environments or those with long development times, such as heavy machinery. In environments where business factors are pushing for rapid change in product development, Agile methodologies are more likely to be introduced. And where Agile exists in a business environment, it is also clear that DITA is more likely to be used to support documentation efforts.

There are many key factors that make DITA-based technical documentation complementary to Agile-based product development. This first is that the topic-based approach in DITA assists with incremental development. One of the tenets of DITA is content reuse, encouraging technical writers to 'write once, use many'. This also means that there is no need to re-write what already exists – a writer can simply reuse entire topics, paragraphs, or phrases used elsewhere thanks to reuse mechanisms within DITA. This enables writers to easily keep up with the rapid pace of development changes.

Agile user stories map well to the task topic type in DITA. Scrum-based Agile often calls upon user stories to help craft development efforts.

These often take the form of various procedures that users will want to accomplish. This format fits nicely with the DITA task topic type. One common practice is for the DITA technical writers to encapsulate the concept as the context for a task instead of writing separate concept and task topics. Additionally, Agile 'epics' are collections of related user stories that comprise the complete workflow for a type of user. From a DITA standpoint, epics can be used to help refine audience-based conditional processing of content, or maps (chapters) within a bookmap when an epic story hierarchy exists.

DITA 1.3 also adds troubleshooting as a new topic type, designed to provide specific solutions to scenarios that are likely to arise, and how to solve them. This new DITA topic type is perfect for writers looking for a troubleshooting option for user stories where a task may not be an appropriate solution.

DITA best practices advocate that content is focused squarely on the user, which is the same focus and this ties in nicely with Agile's similar focus. Technical writers are able to provide early feedback on products through their active use of the product. In this way, technical writers often become an advocate for users; this in turn helps define realistic user stories. The constant change and iterations of content over multiple releases force a change in the typical writer's mindset from 'document everything' to instead 'document only what the user needs'. Again, the granular, topic-based nature of DITA helps make this possible.

Individual DITA topics can be counted, allowing for documentation project measurement. In a typical Scrum-based Agile environment, everyone involved in a project gathers together to discuss progress in regular meetings. Using a traditional DTP-based approach, all that can typically be reported is the word count or the number of chapters completed since the last sprint. With DITA, it is possible to match development features to individual topics, making it easier to report in a more realistic manner on documentation progress. If a CCMS is used, workflow status – draft, in review, done, and so on – can also be measured and reported at the Scrum meeting.

The DITA best practice of minimalist writing reduces 'waste' from a customer perspective, which is also in line with Agile practice. One of the key concepts of Lean Management – an Agile methodology – is to reduce waste wherever possible. This is encapsulated in the Japanese term 'muda'. In the case of documentation, this refers to content that is unnecessary in order for the customer to use the product. It serves as a check on technical writers from writing 'filler' – typically background or marketing-related content that a user does not need in order to accomplish a particular task or action with the product. One of the philosophical underpinnings of DITA is minimalism, which similarly tells writers to pare content down to its essentials.

The various content reuse mechanisms in DITA – topic, conrefs, and keys – contribute to greater consistency in documentation output generated within an Agile environment. Due to short deadlines and time constraints within Agile environments, it is more expedient for technical writers to reuse content where it exists, and DITA provides easy mechanisms for accomplishing this. In many business environments, if you have a topic and it has been reviewed and approved, and you want to reuse it elsewhere, it does not have to be sent out for review again, as it has already been done. (The exceptions are regulated environments where all content must be reviewed in context, but even here content reuse speeds up the approvals process).

Thanks to the separation of content formatting built into DITA, technical writers can focus on creating content rather than formatting it. This can save a considerable amount of time. An informal survey I did several years ago with a team of technical writers using a popular DTP software to produce their documentation showed that roughly half of their time was spent formatting content. That time can now instead be put towards writing more Agile content in a DITA-based environment. This also eliminates the time wasted when SMEs comment on formatting instead of the content they are supposed to be reviewing.

Agile encourages continuous feedback and because DITA topics tend to be small, topic-based review is easier for developers than having to review a full chapter or more. In this way, documentation can also support broader communication between teams, customers, audit processes, and so on. Work cycles are faster, and documentation feedback becomes more critical.

The DITA Open Toolkit is designed to produce documentation outputs in multiple formats, including HTML, WebHelp and PDF, so DITA makes publishing on-demand to multiple formats straightforward. It is also possible to flexibly produce documentation at the chapter or individual topic level as needed. Waiting for documentation deliverables is rarely a bottleneck in a DITA-based process.

All of these factors make using DITA within an Agile development environment an ideal choice for documenting content.

The future of DITA

What goes into the makeup of any DITA standard is decided upon by the members of the OASIS DITA Technical Committee (DITA TC). The committee comprises OASIS members who typically have some expertise with the architecture and implementation of DITA. The members are mainly consultants who work with clients who use DITA, representatives from firms with substantial DITA implementations, and from software firms who create tools for use with DITA. The development of a specification is a lengthy process, with a five-year span separating the release of the latest version of DITA over its previous incarnation.

The most recent version of the DITA specification is version 1.3, which was released in December 2015 (OASIS, 2015). This new version included several new features, including the troubleshooting topic type, key scoping, branch filtering and a number of smaller changes over the previous version (OASIS, 2015). While some of the more popular XML editing tools were quick on the uptake to support the updated standard, as of late 2016 the major CCMSs are only beginning to support the full range of features that are available in DITA 1.3. This has delayed the uptake of DITA 1.3 features within the DITA community as a whole. Over time it is expected that most technical writing teams using DITA will move to the latest version of the standard shortly after the specific software toolchain they use fully supports its features.

DITA 2.0

The next major revision of DITA is slated to be DITA 2.0. It is still too early to define any significant changes, but a couple of things are known: DITA 2.0 will be designed so that it is not backwards compatible with DITA 1.x, and that effort will be put into further optimizing architectural features contained within the standard (Eberlein, 2015). The committee intends to provide guidelines to help documentation teams transition from DITA 1.x to DITA 2.0 (Eberlein, 2015), and some of the features that may be re-designed include how chunked content works, improvements to the metadata mechanisms, increased modularity and possibly the removal of little-used features (Eberlein, 2015). DITA 2.0 is likely to be several years in coming; much of the effort of the DITA TC during 2016 was spent on producing an official erratum for DITA 1.3.

Lightweight DITA

In the meantime, the thing to watch is Lightweight DITA, which is likely to debut in some form during 2017. It is actively being worked on by the members of a sub-committee of the official DITA TC, as two of its design goals are to simplify the existing DITA model and to have it no longer rely exclusively on XML-based semantics (Priestley, 2016), freeing DITA for use in other data models such HTML5 and Markdown.

One of the criticisms often levelled at DITA is that it is too complicated for occasional authors to easily use. Lightweight DITA seeks to strip down the DITA element set to a bare minimum. As an example of how 'complicated' full DITA can be, when someone is authoring content within a paragraph, no less than 60 elements are available to further describe that content. A technical writer with experience of DITA will know the difference and contexts in which to use the msgnum, keyword, apiname, foreign and b elements within a paragraph, but a more casual user may experience decision paralysis when confronted with so many different options. Lightweight DITA aims to strip this down to 10 or fewer elements (Eberlein, 2015), many of which have equivalents in HTML, making it much easier to learn. One of the main audiences for Lightweight DITA are subject matter experts within a company who are expected to contribute content to technical publications, who no longer have to learn full DITA.

Many software developers who are recent graduates have little to no direct experience with XML constructs and are more familiar with other formats, such as Markdown and HTML. Markdown is a simplified markup language that uses characters like dashes, octothorpes, asterisks, underlines and white spaces to indicate simple formatting (Wikipedia, n.d.). For example, a heading is commonly indicated in Markdown using a single octothorpe ('#') at the beginning of a short line of text, while two octothorpes ('##') indicates a sub-header. In addition to being used by programmers, Markdown has also found a significant audience for those who create blog posts and short articles for the web, using popular tools such as WordPress (Gruenbaum, 2015). Similarly, many people are already familiar with HTML elements, but do not want to learn DITA. One of the more common scenarios that documentation teams encounter is having to get content from developers who are creating the features of a product, so anything that would make it easier for a subject matter expert to provide content would be welcome. One of the goals of Lightweight DITA is to make it possible for people to write valid DITA content in Markdown (and also in HTML) that can be transformed to XML-based Lightweight DITA. Discussions are still ongoing, but what is clear is that some forms of content reuse will need to be available in the HTML and Markdown formats to make this possible. If the Lightweight DITA proposal comes to pass, DITA will begin to move beyond its roots in XML, opening up the possibility that it will have a lengthy lifetime as a documentation standard.

References

Aerospace and Defence Industries Association of Europe, n.d.. Welcome To S1000D. [Online] Available at: http://public.s1000d.org/Pages/Home.aspx [Accessed 17 February 2017].

Anderson, R. D. et al., n.d., DITA Open Toolkit: Processing Structure. [Online] Available at: http://www.dita-ot.org/2.4/dev_ref/processing-structure.html [Accessed 17 February 2017].

Anderson, R. D. et al., n.d., DITA-OT transformations (output formats). [Online] Available at: http://www.dita-ot.org/2.4/user-guide/AvailableTransforms.html [Accessed 17 February 2017].

Beck, K. et al., 2001. Manifesto for Agile Software Development. [Online] Available at: http://agilemanifesto.org/ [Accessed 17 February 2017].

CDG, 2016. S1000D – What is it? – S1000D 101. [Online] Available at: http://www.s1000d.net/#101 [Accessed 17 February 2017].

CDG, 2016. S1000D – What is it? – Why S1000D Matters. [Online] Available at: http://www.s1000d.net/#matters [Accessed 17 February 2017].

DITAWriter, 2016. Companies Using DITA. [Online] Available at: http://www.ditawriter.com/companies-using-dita/ [Accessed 17 February 2017].

DITAWriter, 2016. Don Day and Michael Priestly on the Beginnings of DITA: Part 2. [Online] Available at: http://www.ditawriter.com/don-day-and-michael-priestly-on-the-beginnings-of-dita-part-2/ [Accessed 17 February 2017].

Eberlein, K. J., 2015. DITA: Past, Present, and Future. [Online] Available at: http://www.slideshare.net/InfoDevWorld/the-past-and-future-of-dita-with-kristen-james-eberlein/28 [Accessed 17 February 2017].

Eberlein, K. J., 2015. DITA: Past, Present, and Future. [Online] Available at: http://www.slideshare.net/InfoDevWorld/the-past-and-future-of-dita-with-kristen-james-eberlein/29 [Accessed 17 February 2017].

Eberlein, K. J., 2015. DITA: Past, Present, and Future. [Online] Available at: http://www.slideshare.net/InfoDevWorld/the-past-and-future-of-dita-with-kristen-james-eberlein/30 [Accessed 17 February 2017].

Eberlein, K. J., 2015. DITA: Past, Present, and Future of DITA. [Online] Available at: http://www.slideshare.net/InfoDevWorld/the-past-and-future-of-dita-with-kristen-james-eberlein/30 [Accessed 17 February 2017].

Gruenbaum, P., 2015. ProgrammableWeb: Why You Should Use Markdown for Your API Documentation. [Online] Available at: https://www.programmableweb.com/news/why-you-should-use-markdown-your-api-documentation/2015/02/19 [Accessed 17 February 2017].

LinkedIn, n.d.. S1000D – DITA bridge over troubled water. [Online] Available at: https://www.linkedin.com/groups/7406011 [Accessed 17 February 2017].

OASIS, 2002. The Simplified DocBook Document Type. [Online] Available at: http://www.oasis-open.org/docbook/specs/cs-docbook-simple-1.0.html [Accessed 17 February 2017].

OASIS, 2015. #DITA V1.3 OASIS Standard published. [Online] Available at: https://www.oasis-open.org/news/announcements/dita-v1-3-oasis-standard-published [Accessed 17 February 2017].

OASIS, 2015. B.2.1 Changes from DITA 1.2 to DITA 1.3. [Online] Available at: http://docs.oasis-open.org/dita/dita/v1.3/os/part3-all-inclusive/non-normative/new-in-1.3.html [Accessed 17 February 2017].

OASIS, 2015. DocBook Version 5.1: Committee Specification Draft 01 – 2.1. Assemblies. [Online] Available at: http://docs.oasis-open.org/docbook/docbook/v5.1/csprd01/docbook-v5.1-csprd01.html#s.assembly [Accessed 17 February 2017].

Priestley, M., 2001. DITA XML: A Reuse by Reference Architecture for Technical Documentation. [Online] Available at: https://pdfs.semanticscholar.org/fa3c/f7617c3c1ff4bb937f97d9a2b409036a32 57.pdf [Accessed 17 February 2017].

Priestley, M., 2016. Lightweight DITA: A pre/overview. [Online] Available at: http://www.slideshare.net/mpriestley/lightweight-dita-a-preoverview/6 [Accessed 17 February 2017].

Walsh, N., 2016. What is DocBook?. [Online] Available at: http://docbook.org/whatis [Accessed 17 February 2017].

Wikipedia, n.d.. Wikipedia: Markdown. [Online] Available at: https://en.wikipedia.org/wiki/Markdown [Accessed 17 February 2017].

Wikipedia, n.d.. XML – History. [Online] Available at: https://en.wikipedia.org/wiki/XML#History [Accessed 17 February 2017].

York, D., 2001. Chasing Documentation's Holy Grail: Introduction – DocBook History. [Online] Available at: http://lodestar2.com/people/dyork/talks/2001/xugo/docbook/frames/foil04.h tml [Accessed 17 February 2017].

11

Automatic documentation for software

Andrew McFarland Campbell

Abstract

Many software products are complex and have anything from a dozen to thousands of configuration parameters. This is particularly true for distributed or cloud-based systems. Traditional documentation methods have trouble keeping up with development, especially in Agile development environments where there may be many subtly different product versions.

It is increasingly important to automate as much of the documentation process as possible, as this reduces the burden of effort required to create documentation, and because it ensures that documentation is always up to date. It also makes it much easier to produce different versions of the documentation to match the different software versions.

Keywords   automation; Agile; Javadoc; Swagger

Introduction

Over the past decade I have been involved in a wide range of projects where automatic documentation generation has been used. Probably the most challenging aspect of automatic documentation is changing your mindset from *writer* to *developer*. Automatic documentation requires you to interact with and possibly change the source code. It can also mean that developers become part-time writers. In this chapter, I give an overview of my experiences.

Standard solutions

Every software developer knows that, wherever possible, you should use a solution that already exists. The same is true for automatic documentation. Your first approach should be to see what current tools exist before developing your own.

In one project, I was documenting a Java software development kit (SDK) aimed at experienced Java developers. The natural choice for documentation was Javadoc, a mature documentation generator for Java (see 'Suggestions for further reading' on page 235). Javadoc takes specially formatted comments in the source code and converts them into documentation in a standard format. Most Java SDKs and libraries have documentation written in Javadoc, and the format of the output is familiar to most Java developers. It made sense for this project to use Javadoc because the final documentation deliverable would be in a format that the end user was already familiar with.

It also made sense because the SDK developers were themselves familiar with how to write Javadoc comments in their code. There was a very shallow learning curve for both the developers and the end users.

```
/**
* Instantiates a new McFarlandCampbellObject
* with a colour and message
* @param colour The colour
* @param message The message
*/
    public McFarlandCampbellObject(final String colour, final String messsage )
{
// java code here
}
```

Figure 1: Some simple Javadoc

McFarlandCampbellObject

public McFarlandCampbellObject(String colour,
                    String message)
Instantiates a new McFarlandCampbellObject with a colour and a message

Parameters:
colour - The colour

message - The message

Figure 2: The resulting documentation

Javadoc can be configured to warn if the documentation is incomplete (if, for example, a parameter isn't documented). That means if Javadoc runs without error or warning, then the code has complete documentation coverage. However, full coverage of the code is only half the battle. That coverage must make sense as well. Consider this example:

```
/**
 * Document this later!!!!
 * @param colour a param
 * @param message a param
 */
    public McFarlandCampbellObject(final String colour, final String messsage)
{
// java code here
}
```

Figure 3: Incomplete Javadoc

That code will pass through the Javadoc generator without any errors or warnings, yet it clearly doesn't document the software. When working with developers who add Javadoc to their code, rather than acting as a writer, you act as an editor. Your job is no longer creating content: your job is making sure that content that other people create is accurate and useful.

Most technical writers will have acted as editors at some point in their careers. Editing content written by developers is very different to editing content written by other technical writers. Substantial changes may be needed, and this is where your people skills as a technical writer come into use. I have worked with some very talented developers who can write world-class code, but who write poor English; sometimes this is an issue, not for me, but for the programmers. It is very easy to unintentionally embarrass developers over what they have written and how they have written it. If a developer is embarrassed by a change I make to their English, I usually say something like "Don't worry, you might be worried about your punctuation, but you should see my Java!"

The biggest problem with Javadoc and similar systems is the false sense of security that they can give you. We have already seen that just because the documentation generator completes without any errors or warnings that doesn't mean the documentation is useful. Even if you have meticulously reviewed every Javadoc line, and made sure that all of the documentation is complete, you may not actually have useful documentation. Complete Javadoc will tell you how to use class and method, but it may not tell you how to use those classes and methods in a practical situation. Additional documentation, such as developer guides or tutorials, may be necessary, depending on your target audience.

When additional documentation is necessary, you want to keep it as closely coupled to the code as you can. Even if the additional documentation is completely manually written, as it probably will be, keeping it in the same source-code repository as the code itself is always a good idea. That way your documentation benefits from all the branching and versioning that the code itself does. If there is a feature that is only in branch 2.2.3, then if you are using the same source-code repository the documentation for that feature will only be in the documentation for branch 2.2.3. I have often found that technical writers are resistant to branching documentation like this, partly because of the perceived complexity of the branching and merging process, but also because of the belief that end users will be confused by multiple versions of the documentation. As an end user, I prefer to have documentation that is specific to the version of the software that I am using, because it means I don't have to remember which features are available in the version I am using. If branching and merging are used correctly – the core or trunk of the software is fully documented, and new features are documented on the branches that introduce them – then the documentation will be as complete as the software.

When writing supplemental documentation like this, I like using DocBook or DITA. That way the final documentation output can be generated at the same time that the final product is built. DocBook and DITA are, of course, XML so your documentation can further benefit from all the branching and merging that the source-code repository can provide. The alternative solution, where multiple binary-based documents, such as Word documents, are managed independently of the source code, can become extremely time-consuming and error-prone to maintain.

Solutions in other languages

Javadoc works for Java. Other languages have similar solutions. For example, there is a tool called Doxygen (see 'Suggestions for further reading' on page 235) that can provide Javadoc-type documentation for a wide range of languages, including C, Objective-C, C#, PHP, Java, and Python. When using Doxygen, or whatever existing solution you choose, the same concerns and caveats apply: simply because the documentation compiles without warning or error does not mean that the documentation is complete or helpful. Additional documentation is almost always needed. The important benefit that automatic documentation provides is that instead of spending many hours maintaining, say, API reference documentation, you can spend those hours writing user guides and tutorials that can't be generated automatically.

Customized solutions

There will be projects where there is no suitable existing solution. Every software developer knows that when there is no existing solution to a problem, you have to create one. For example, I once worked on a project for a system where source data was manipulated through several intermediate stages before being processed into a final output.

Using Doxygen, it would have been fairly simple to produce documentation that mapped between the input format and the first intermediate format, and then from the first intermediate format to the next, and so on, until the final output was reached. The end user could then have followed through the documentation to determine how the input was transformed into the output. This would have been unsatisfactory for two reasons:

- Working through the documentation like this would have been extremely frustrating for the end user
- The documentation would have exposed a lot of the detail of the implementation, which would mean that internal changes that should be invisible to the end user would have been exposed.

I wrote a custom solution that took each stage of the mapping and condensed it down to show just the input-to-final mapping. This was much easier for the end user to use, as well as keeping the internal workings of the software hidden.

For the technical writer, however, this was much more challenging. I had to create a program that parsed the source code and assembled the documentation. That is obviously a programming task with all the complexities that that involves, such as testing and bug tracking. The end result was worth it though. The end user had a five-page document that told them everything they needed to know, rather than a hundred-and-five page document that told them more than they needed.

On another project I had to produce documentation based on configuration files. In this case, the default configuration files had every possible configuration parameter, with the default values explicitly set, and a comment describing the parameter. A fairly simple script was used to convert the configuration file into a DocBook document, which was then in turn included in the main 'hand-written' documentation.

Although these custom solutions were very different, they both had something in common. They both produced DocBook as output, and then that DocBook was processed using the standard DocBook tools. That is because, even when working with a custom solution, you want to use as much standard technology as you can. Could I have written a script that would have parsed the configuration file and turned it into HTML? Yes, that would have been as easy as the one that produced DocBook. Could I have written a script that could have turned it into PDF? Yes, but that would have been considerably more complex. By choosing DocBook as the output format, I got all the advantages of using a robust, existing tool chain. When creating a custom solution, you want to keep the amount of custom code to an absolute minimum.

Swagger

No discussion of automatic software documentation would be complete without mentioning Swagger (see 'Suggestions for further reading' on page 235). Swagger is a language that can be used to define and describe RESTful APIs. A RESTful API is one sort of API for web services, and RESTful APIs are becoming very common, particularly in cloud computing. Swagger can be written in either JSON or YAML format, and if the Swagger is correctly written, it tells you everything you need to know about your RESTful API.

Figure 4 on page 231 is an extremely simple example for a RESTful API with only one endpoint. For more detailed Swagger examples, visit http://www.swagger.io.

There are two ways that Swagger can be used. The first is the top-down approach, where someone creates a specification in Swagger, and then develops the API to match the specification, using some of the tools that can convert Swagger into skeleton code. The second is the bottom-up approach, where you have an existing RESTful API and you create Swagger to describe it. This approach can be done automatically by using specially formatted comments in the code, in a way not dissimilar to Javadoc.

```
{
    "swagger": "2.0",
    "info": {
        "version": "1.0",
        "title": "Simple Swagger Example"
    },
    "paths": {
        "/servicestatus": {
            "get": {
                "responses": {
                    "200": {
                        "description": "Status returned when everything's OK"
                    }
                }
            }
        }
    }
}
```

Figure 4: Simple JSON Swagger example

As technical writers, we aren't too concerned with how the Swagger is generated or used by the developers. The thing that makes Swagger useful for us is that the Swagger completely describes the API, and does it in a structured way. Swagger contains all the information you need to create end-user documentation.

You could read through the Swagger file and manually update your documentation in FrameMaker, but I hope you can see that that would be somewhat impractical, and against the spirit of this chapter. You might think that you should take the Swagger and convert it into DocBook or DITA using some standard tool, such as Swagger2Markup (see 'Suggestions for further reading' on page 235). While that is an improvement, there is a much more exciting way to use Swagger in technical writing.

There is a tool, called Swagger UI (see Figure on page 235), which is simply a collection of HTML and JavaScript that runs in a standard web browser. Swagger UI takes a Swagger file as input, and parses it to produce an interactive web page that not only documents the API, but it allows you to interact with the API itself. You can enter different parameters into the fields on the web page, send the request directly to the API, and then the response is displayed. With Swagger UI, the documentation isn't just static documentation. It is an interactive portal for hands-on learning.

Non-technical issues

In this chapter I haven't written very much detail about the specific technical aspects of automatic documentation. That's partly because technology develops so quickly that any technical detail would quickly become out of date, and it is partly because to thoroughly cover just one of the technologies I have mentioned could take more space than we have in this book, let alone this chapter.

The main reason is because the technical issues are the easy ones. The hard issues are the human issues. For automatic documentation to work, you have to get writers to think like programmers, you have to get programmers to think like writers, and you have to get management buy-in.

Writers thinking like programmers

For a technical writer who is used to creating traditional documentation, making changes in the source code can be an intimidating experience. Source code is usually edited in a text editor or in an integrated development environment (IDE). Even though these environments generally don't have hard learning curves, they are very different to working in Word or FrameMaker. Before a writer can be involved in an automatic documentation project, they have to be comfortable editing source code.

I have been developing code for the best part of 20 years. I'm familiar with changing source code. Even if I am working with a programming language I have never encountered before, I can make an educated guess about which parts of the code I can change (for documentation purposes) without breaking anything. Someone who has only ever written in Word or an XML editor lacks that experience. How do you get that experience? You get it by experimenting with the source code. Get a copy of the project, change a few comments and compile things to see if it breaks. When it breaks, see if you can change things back until you find what broke it. If you don't know how to compile the code, or if you can't understand the error messages from the compiler, then speak to one of your programmer colleagues.

When technical writers start to use the source-control system, with branching, and versioning, and tagging, they also start to experience the benefits of it. If you break something, and the code no longer compiles, and you can't see what you broke, you can simply reverse the changes from source control. Once you realize you can easily reverse changes you have made, you will become much more confident when you are making changes, because the worst-case scenario is an easily fixable one.

If all of this sounds like technical writers becoming junior programmers, it is because they are. The technical writers aren't making changes to the functionality of the final program, but they are interacting with the source code in the same way that the developers do.

Making programmers think like writers

In the traditional model of software documentation (where developers write code and the technical writers create their documentation in related but distinct processes) the comments that developers write in code are never exposed to the end user: at worst they are read by other developers or technical writers. When working with automatic documentation, this is no longer the case. A programmer can write a comment in code which will end up, unedited, in the documentation.

Developers have to be aware that when they write a comment (in a format that the documentation system will pick up) then that comment will be seen by the end user. Just as some technical writers can be intimidated by the thought of editing source code, some developers can be terrified at the thought of writing and editing prose. This is where good team communication is essential. A technical writer who can't work out how to get the code and documentation to compile should be able to speak to a developer colleague, and a developer who isn't sure of the rules of punctuation should be able to speak freely to a technical writer colleague.

As technical writers, we are used to having our prose edited. It doesn't offend us, it doesn't make us uncomfortable. That is not always the case with non-writers. Sometimes people get very attached to particular phrases, or particular punctuation styles. Perhaps the most difficult part of automatic documentation for developers is realizing that their prose will be changed. Just as technical writers have to think like developers, the developers have to learn to think like technical writers.

Management buy-in

The most important aspect of any change to the software development lifecycle, such as introducing automatic documentation, is getting management buy-in. If you can convince your management that automatic documentation is the right thing for your project, then you will be able to go ahead with it.

What can you say to convince management to allow you to introduce an automatic documentation system?

Automatic documentation will result in better quality documentation, and ultimately automatic documentation will require fewer resources.

The first of those points I have mentioned before. An automatic documentation system will mean that your reference documentation will be much more up-to-date than manual documentation. A new parameter is added to the code, it is automatically in the documentation. It doesn't matter that that particular sprint was completed when the tech writer was on holiday, the documentation is always up to date.

Automatic documentation will require fewer resources. You don't need a full-time technical writer to manually update your reference documentation. You just need someone who can spend a little time proofreading the changes made by the developers.

That doesn't mean that you can fire half your technical writing team. It does mean that all those technical-writer hours that previously were used on manual documentation can be used for other documentation, the sort of things that can't be automated, like procedures and tutorials. Ultimately, that means that the finished product has more thorough documentation, making it more valuable.

Management has a crucial role in automatic documentation. When automatic documentation requires particular coding or commenting styles, those styles must be enforced by management. For example, if a developer writes incomplete javadoc, as in Figure 3, management must require that this is fixed before the code can pass code review. Following the correct processes throughout the development lifecycle is more efficient that trying to fix incomplete comments at the end of the process, and comments must never be seen as 'cluttering' the code.

Conclusion

For me, the most compelling reason to automate documentation is a philosophical one. Every technical writer knows about the importance of separating content from presentation. When you generate documentation from the source code, whether you are doing it with a simple solution like Javadoc or you are doing it with a custom system that parses the source code, you are completely separating the content from the presentation. The content isn't a file that describes the program, the content is the program itself; the documentation is, in effect, an alternative view of the compiled software. This makes the final documentation much better for the end user, and, after all, that is what technical writing is all about.

Suggestions for further reading

Javadoc Tool Home Page:
http://www.oracle.com/technetwork/java/javase/documentation/index-jsp-135444.html

Doxygen: http://www.stack.nl/~dimitri/doxygen/

Swagger: http://swagger.io/

Swagger2Markup: https://github.com/Swagger2Markup

Swagger UI demo: http://petstore.swagger.io/

Part 3

Roles of Technical Communicators

12

Trends in technical communication in Ireland

Yvonne Cleary

Abstract   Technical communication is a new field in Ireland, and most technical communicators work in software environments. Worldwide, technology and globalization have altered technical communication practice considerably. This chapter reports on the findings from interviews with nine technical communication practitioners, managers and company owners, all based in Ireland. They discussed trends affecting the field, predicted future trends and made recommendations for curriculum content. The findings indicate that practice is impacted most by new media, Agile development, virtual teams and off-shoring. Foundational skills such as writing and information design should remain prominent, as technical communication curricula develop to respond to trends.

Keywords   technical communication; Ireland; new media; Agile; virtual teams

Introduction

Spilka (2010, p. 2) described the extraordinary impact of the digital revolution on technical communication practice worldwide, noting how it has altered every aspect of the technical communicator's job, including how they "gather information, think, develop and share ideas, collaborate, analyze, plan, and make decisions; find, use, manipulate, manage, and store information; and develop, evaluate, revise, and complete information products."

In Ireland, as elsewhere, technical communication practice is affected by technological, economic, corporate and social developments. Understanding trends in an occupational field and being able to distinguish between long- and short-term trends are important strategic and professional activities. Strategic knowledge about potential developments in, and directions for, the field enables practitioners and managers to prepare for the future. This knowledge is also important to academics to ensure that academic programmes are appropriately designed and equipped to train students for their future careers.

In this chapter, I discuss the findings from interviews with technical communication practitioners, managers and company owners about their perspectives on the trends that are shaping the discipline. In Ireland, technical communication practice is concentrated in software, rather than manufacturing, industries. Therefore, trends in technical communication for software industries are the focus of the chapter.

This chapter begins with an overview of the recent literature about trends in technical communication, it then describes the interview methodology and participants, next it discusses the interview findings, and finally it outlines how these findings inform the teaching of technical communication.

Trends affecting the field of technical communication

Globalization and technology have had a remarkable impact on the field and practice of technical communication in the past two decades. Technology has transformed work and the practice of workers in almost all disciplines. Myriad modes of collaboration, previously unimaginable, are now routine. Technical communicators need to be able to explore and expand their digital literacy, in order to operate effectively in this milieu. A recent study by Dubinsky (2015) shows that while managers believe the role needs to expand, they are not in agreement as to how. Keeping abreast of technology trends and software for the field are essential competencies for technical communicators. Brumberger and Lauer's (2015) analysis of over 900 technical communication and related job advertisements over a two-month period shows that technology, and software competence in particular, is a focal point in job advertisements, and hybrid and digital roles are prominent.

Trends in content interact with and are influenced by technology developments. Recent publishing phenomena driven by social media, and encompassing products such as wikis, podcasts, discussion forums, video, and blogs, have utterly changed content production processes. In an era of YouTube, Twitter, blogs, smart phone applications, ebooks, embedded assistance, and myriad other instructional content applications, almost anyone anywhere can publish and access information. Nevertheless, some social media content is shoddily produced, and "unprofessional", leading to the phenomenon Keen (2007) describes as the "cult of the amateur". A related development, 'crowdsourcing', sees the 'crowd' (comprising any online user who wishes to contribute) creating and contributing to content and products, usually without receiving payment for their contributions. Crowdsourcing is used extensively by corporations to support products, especially for localization and customer support tasks. Crowdsourcing may be a challenge for practicing technical communicators, who may see their role usurped if support content is developed for free. Nevertheless, the abundance of information now widely available is also a potential opportunity for technical communicators, often called information professionals, who are skilled in designing content that is easy to access and use. One opportunity for writers, for example, is to move towards roles involving content strategy and content curation. Hart-Davidson (2001, p. 145) explained the increasing importance of technical communication in the information age. "[M]ore and more, the exchange value of an information product is associated with aspects of quality that technical communicators have the expertise to look after: customization for specialized or niche audiences, ease of use, and scalability." This comment remains apposite in an era of socially created content.

Work practices also impact on technical communicators. The practices of outsourcing and off-shoring may pose challenges, for example. These phenomena are driven by globalization, and are a product of the information age. Paretti, McNair and Holloway-Attaway (2007, pp. 327–328) argue that "outsourcing, offshoring, and globalization, enabled by a dynamic network of communication technologies, have altered the physical and social landscapes of our working lives." Another work practice that changes how technical communicators work is the adoption of Agile methodologies in software development. 'Agile' development environments – where products and associated documentation are developed incrementally, rapidly and collaboratively in 'scrums' – represent a new but increasingly prevalent writing paradigm for technical communicators (Dubinsky, 2015). Structured authoring standards such as XML and DITA have also changed work processes (Mott and Ford, 2007).

Giammona (2004) discussed the future of technical communication following a study involving people "known to be thought leaders" in the profession. She posed the question "Will we disappear?" and she referred to the STC 50th conference where the concluding presentations all made dire predictions for the future of the field. In addition, many of her survey respondents were downbeat about the perceived importance of technical communication. Because the technical communication field, and technology itself, is in flux, it is an interesting time to examine practitioners' views of the trends in their field.

Methodology

This chapter draws on data from a larger study (see Cleary 2016) that included interviews with nine Irish technical communication practitioners, managers and company owners based in Ireland. The interviews explored respondents' views on trends in technical communication workplaces, practices and education. The respondents also commented on the impact of economic trends on the field. I also asked them to recommend subjects they believe should be included in future technical communication curricula. The interviewees were selected as a purposive sample. The following table presents demographic and professional data about the interviewees. To ensure anonymity in reporting the findings, interviewees are given pseudonyms and I have not revealed company names.

Table 1: Interviewee details

Pseudonym	Job title	Gender	Employer
John	Technical communication team manager	Male	Large multinational with a team of technical communicators
Tony	Technical communicator	Male	Large multinational with a team of technical communicators
George	Company owner	Male	Small technical communication company
Patrick	Company owner	Male	Small technical communication company
Mary	Lone writer	Female	Lone writer in a small online research company
Alice	Technical writer	Female	Small software company
James	Contractor	Male	Self-employed

Table 1: Interviewee details (continued)

Pseudonym	Job title	Gender	Employer
Regina	Technical communication team manager	Female	Large multinational with a team of technical communicators
Jane	Information developer	Female	Large multinational with a team of technical communicators

Considering the small sample, and the diversity among respondents, the findings cannot be generalized to the larger population of Irish technical communicators. The purpose of the interviews, rather, was to gain a multiplicity of perspectives.

Interview findings

The following sub-sections explore the key trends that emerged from the interviews: writing trends, Agile work environments, virtual teamwork, and off-shoring and outsourcing.

Writing trends

A trend that all interviewees predicted into the future is the increasing impact of technology on writing. This trend has several facets, including automation of some writing and design tasks, the need for more emphasis on content strategy, lean content, structured authoring (and particularly DITA among these interviewees), and the movement of more services and documentation online.

While technologies are pivotal to technical communication practice, the rapidly changing technological sphere and consequent range of tools and techniques that writers must master pose challenges for practice. New media, and especially social media tools, have the power to expedite and automate content development and publication. The interviewees were somewhat divided on the impact of new media on technical communication. For example, John suggested that user-generated content can give writers valuable feedback. "If a customer has a problem they just hit the feedback button and the feedback goes straight to the author." Moreover, he argued that user-generated content: "gives vibrancy to your product. That kind of life is invaluable, people talking about it, working on it." Conversely, other interviewees saw new media as a threat to technical communication. George considered cheap online vendors to be a potential threat. Mary suggested that social networking sites such as Facebook will reduce the need for dedicated websites and by extension, web designers.

> Social media is a big threat. Companies might eventually not have websites but use Facebook outposts. There's a lot more information moving on to Facebook now, and people don't go to your website anymore; they go to your Facebook page.

These perspectives are also reflected in the literature on the use of new media in technical communication, which suggests that technologies that enable users to contribute give writers good feedback about their work (Gentle, 2010), but new media content generated by users is often of poor quality (Keen, 2007; Howe, 2008), and benefits from editing and content curation by specialists (Gentle, 2010).

The increasing importance of structured authoring, single sourcing, and use of DITA and XML in writing is reflected in these interviews. Interviewees saw DITA and XML as growth areas and required skills in the field, though they reported that DITA is much more likely to be implemented in large writing teams. Many interviewees also highlighted the shift in the role towards managing and curating content, and towards producing different types of content (video, interactive online content, interface messages) rather than only writing manuals. A related writing trend that interviewees discussed is lean documentation. Dubinsky's (2015) research with managers indicates that "lean" content is a dominant trend.

Agile work environments

Some of the companies for which interviewees worked were in the process of adopting an Agile development model. The managers and writers I interviewed indicated that this transition was proving challenging, but should ultimately increase the visibility of the writing team. Regina explained how the Agile model works:

> All the authors are assigned to scrums, the software is developed in sprints, with maybe 10 or 12 sprints to a release. Each scrum will have one manager assigned to it. You hear mixed reports but overall it's been very positive. For the first time writers are now very much part of the team. It's easier to get yourself on the agenda. You're there. You're planned in. It's all visible. So that's really positive.

Tony agreed that Agile environments increase visibility:

> In an Agile team, you have writers, developers, and so on, but they're supported by... [other functions], who would be supporting a range of teams. The [writers] are not sidelined in the organization structure. Relationships are very good.

Findings from Dubinsky's (2015) study with managers report a similar outcome of Agile, in integrating the writer into the development process.

For lone writers in smaller software companies, Agile also appears to be the prevailing development model. The quote below, from a lone writer, indicates that while she was part of the development team, her role was not clearly defined, and she benefitted most by gathering information during meetings.

> I work with an Agile team, but [am] still working out the role of the writer. I'm not a formal part of the team [...] I participate in meetings, but mainly as a listener so I know of documentation issues that may come up. It's useful in getting to know the team, and also informally raising or answering documentation issues. The experience has been positive.

However, the transition period can be challenging. John noted how the move to Agile "is causing a lot of pain for a lot of teams." One difficulty of the Agile model for managers, especially in the early stages of adoption, is in assigning the correct number of writers to projects, as illustrated in this quote.

> The software development model we used before was the Waterfall model, and that stipulated the number of developer days, depending on the functionality. Based on those numbers, we would stipulate the number of information developers. We applied that formula to the Agile model. Sometimes it worked; sometimes it wasn't enough time, sometimes it was too much. People would be assigned to two scrums and in reality one scrum might be half what they thought it would be and the other would be twice as much. Experience and maybe a few more probing questions to developers around the beginning of a project will help us [to calculate more accurately].

Virtual teamwork

'Virtual' teams, where members are situated in different geographical locations, are commonplace in large global corporations (Robey, Khoo and Powers, 2000). For Irish technical communicators who work in virtual teams, the day is split into time spent writing, and time spent communicating with team members in other locations. For example, Tony writes in the morning, and uses his afternoons for team meetings. "I work with the States, exclusively. My entire team is in the US. So they wake up at 2[pm]. My day works back, not forward. I've got the morning to get stuff done."

John, who manages teams in three locations, described how time zones dictate the structure of his working day:

> In the morning I tend to deal with China, to get ID plans together for their new releases, reviewing any work, dealing with resource issues. Through the day I deal with Ireland primarily, doing the same thing for those guys. And in the afternoon it's all US, so it's offering team meetings, every senior manager is involved in those meetings.

The structure of Agile environments does not lend itself especially well to virtual teaming, as explained by Regina:

> What's challenging is that we're here in Ireland and the development scrums are usually in Bangalore. We dial in. But that can be challenging because the whole idea of a scrum is that everyone is sitting in one room so you have 10-minute meetings every day as part of the scrum.

Nevertheless, she stated that technology reduces these challenges. "It'll take some time to figure out, but working virtually is getting easier all the time because the technology is getting better and better."

The findings overall suggest that in Ireland, as worldwide, the work environment for technical writers is globalized and subject to the demands, challenges and rewards of globalization.

On the MA in Technical Communication and E-Learning at the University of Limerick, we run a virtual team project whereby our students collaborate on assignments with students from the University of Central Florida. We exploit information and communication technologies (ICTs) to manage the collaboration process. Findings from the interviews suggest that the virtual team experience is a valuable one for students since they are likely to work in virtual teams following graduation.

Off-shoring and outsourcing

Virtual teamwork itself is a consequence of how technology enables work to be 'off-shored'. The findings from these interviews suggest, as others have done (see, for example, Jablonsiki (2005)) that technical communicators are aware of the potential consequences of corporate restructuring. Technology developments enable work to be distributed, and off-shored to cheaper locations. Many Irish technical writers are insecure about Ireland's ability to compete with locations such as China and India, as evidenced by quotes from interviewees:

> Is all technical writing going to move to India in the next 10 years?!

> I guess our biggest threat is that they will set up a team somewhere else like India.

> And there's off-shoring as well. Like anything, quality is always impacted, but they don't do it for quality, they do it for costs. [...] Will tech writers still exist? Yes. Will they be Polish? Probably. We have Polish writers. We have Chinese writers. I always said localization as a function was under threat from off-shoring, but writing wasn't. That's no longer true.

Nevertheless, both managers whom I interviewed saw Ireland as retaining a competitive advantage, primarily because of the base of educated native English speakers. The quotes from John and Regina, respectively, that follow elaborate on this position.

> Off-shoring is a threat but most companies who have off-shored to India and China... an awful lot have pulled back [because of] language quality. In India you've got huge turnover issues as well. China less so. Non-native English speakers will always struggle, unless they're gifted. [...] So I see Ireland as having a huge opportunity right now. Our cost base has come way down, in comparison to Germany, the US, the UK. We're one of the cheapest English-speaking countries to do business in, so we're at a huge advantage.

> Offshoring is always a possibility, but it always has been. Having native English speakers and a suitable timezone

outweigh the cost factor. Let's hope it continues that way. [This company] is definitely hiring in other locations. You do see it going that way but I don't think it will ever move fully. The fact that we have a knowledge economy here and they're so highly qualified means we can hold our own.

The contrasting views of managers and writers on the threat of off-shoring suggest that managers may expect stability, even though experience from other countries, notably the United States, suggests the converse. North American technical communicators have experienced off-shoring since the 1990s, an experience that may be mirrored in Ireland in the coming decade.

Projections for the future

Interviewees were asked to describe the trends that they believe will have most impact on the field of technical communication in the future. Table 2 summarizes their responses.

Table 2: Interviewees' projected future trends

Interviewee	Projected future trends
John	Interface prompts/messages replacing manuals in the end-user software market
	User-generated content , which may be a possible opportunity rather than a threat for writers, because they can be involved in managing that content
Tony	User interface (UI) design, message design, and reduction in the amount of documentation
	More off-shoring of writing to China/India with some need for post-editing
	Technical communication evolving into an editorial function
George	Movement of services online
	Off-shoring technical writing services to cheaper countries
Patrick	New areas for support documentation: engineering support, financial services, training, and policy and procedures

Table 2: Interviewees' projected future trends (continued)

Interviewee	Projected future trends
Mary	Movement of services online resulting in less need for large manuals, and more need for "timely, concise, easily digestible information"
	The threat of social media to traditional customer service interfaces
Alice	Single sourcing
	More use of social media such as wikis to develop documentation Increasing exploitation of user-generated content
James	Increasing move towards online documentation
	Need for writers to certify and market their skills
Regina	The "global nature of software development" and Agile development
	Increasing use of virtual communication tools
	Delivery of information in simpler ways, with less text
Jennifer	Role of the technical writer becoming more technical and less focussed on writing skill and style

As Table 2 illustrates, many interviewees reflected on both the challenges posed by, and the potential to exploit, technological innovations. Technology and globalization have also affected their careers and they expect these trends to continue into the future. Recurring themes are the movement of documentation and services online, leaner documentation, and the possibility of off-shoring. The interviewees also expressed their views on how trends could inform teaching and curricula in technical communication.

Teaching implications

All interviewees commented on the content they would want included in a technical communication curriculum. Table 3 summarizes their responses.

These findings indicate the importance of teaching new media tools, and techniques for new media development and content curation, to technical communication students. The interviewees' recommendations

Table 3: Content for inclusion in a technical communication curriculum

Interviewee	Recommendation
John	Structured authoring and especially DITA
	Internships should be an essential component of programmes to introduce students to the workplace
Tony	Plain English, structured authoring, and communication
	Software for technical communication
George	Interviewing, communication skills (especially the art of extracting information), internships
Patrick	Extensive content on the practice of writing: "writers do not need to be specialists in the technology they are writing about, but they must be language specialists"
Mary	Designing and delivering presentations
	Rhetorical exercises that analyze how users interact with online information
Alice	Information mapping, which feeds into structured authoring and would help students who later work in DITA environments ("in terms of being able to identify well, what is a concept? In terms of reusing content")
James	Mind mapping, more use of software tools, the whole concept of single sourcing: "If there had been a project that we put together and single sourced. I would never have understood that if I had started working immediately"
Regina	Grammar, punctuation and the mechanics of writing
Jennifer	Interpersonal communication and interviewing skills

regarding curriculum development focussed heavily on established skills – grammar and writing, information design and plain English – indicating that while trends are important, foundational skills must also be prioritized. This point supports Carliner's (2010) argument that durable skills must prevail in curriculum development. In the words of one interviewee:

> The world is not getting less complicated, or less regulated. The converse. So there's always going to be a need for people who can explain things. It is the subjects that need explanations that will change. I have an Apple Mac manual from the 1980s that

includes a two-page explanation of how the Backspace key works. These kinds of concepts no longer require procedural explanations, but they are replaced by new information needs. We're not explaining the same things anymore. But we're still explaining things.

Although software was mentioned, interviewees did not recommend specific programs or types of software. Brumberger and Lauer's (2015, pp. 240–241) analysis of job posting shows that employers value tools knowledge and that graduates need "a conceptual understanding of different types of technologies." Two interviewees recommended that internships should be built into programme structure where possible.

Conclusion

The purpose of this chapter was to explore writers' and managers' predictions for their discipline, considering rapid changes due to globalization and technological advances. The interviewees were positive about many of the changes in their field, while acknowledging challenges, especially in relation to off-shoring, and the impact of new media on technical communication practice.

In relation to curriculum development, the findings indicate that it is important for educators to be aware of changes in development practices and outputs and to adapt curricula where possible to take account of these changes. Nevertheless, foundational skills in technical communication must remain prominent.

References

Brumberger, E., and Lauer, C. (2015). 'The evolution of technical communication: An analysis of industry job postings', *Technical Communication*, 62(4), 224–243.

Carliner (2010). 'Computers and technical communication in the 21st century', in Spilka, R. (ed.) *Digital Literacy for Technical Communication: 21st Century Theory and Practice*, New York: Routledge, pp. 21–50.

Cleary, Y. (2016). 'Community of practice and professionalization perspectives on technical communication in Ireland', *IEEE Transactions on Professional Communication*, 59(2), 126–139.

Dubinsky, J. M. (2015). 'Products and processes: Transition from product documentation to... integrated technical content', *Technical Communication*, 62(2), 118–134.

Gentle, A. (2010). *Conversations and Community: The Social Web for Documentation*, Fort Collins, CO: XML Press.

Giammona, B. A. (2004). 'The future of technical communication: how innovation, technology, information management, and other forces are shaping the future of the profession', *Technical Communication*, 51(3), 349–366.

Hart-Davidson, W. (2001). 'On writing, technical communication, and information technology: the core competencies of technical communication', *Technical Communication*, 48(2), 145–155.

Howe, J. (2008). *Crowdsourcing: Why the Power of the Crowd is Driving the Future of Business*, London: RH Business Books.

Jablonski, J. (2005). 'Seeing technical communication from a career perspective: the implications of career theory for technical communication theory, practice, and curriculum design', *Journal of Business and Technical Communication*, 19(1), 5–41.

Keen, A. (2007). *The Cult of the Amateur: How Blogs, MySpace, YouTube, and the Rest of Today's User-generated Media are Destroying Our Economy, Our Culture, and Our Values*, New York: Doubleday.

Mott, R. K. and Ford, F. D. (2007). 'The convergence of technical communication and information architecture: managing single-source objects for contemporary media', *Technical Communication*, 54(1), 27–45.

Paretti, M. C., McNair, L. D. and Holloway-Attaway, L. (2007). 'Teaching technical communication in an era of distributed work: a case study of collaboration between U.S. and Swedish students', *Technical Communication Quarterly*, 16(3), 327–352.

Robey, D., Khoo, H. M., and Powers, C. (2000). 'Situated learning in cross-functional virtual teams', *Technical Communication*, 47(1), 51–66.

Spilka, R. (2010). *Digital Literacy for Technical Communication: 21st Century Theory and Practice*, New York: Routledge.

13

The collaborative effects of cyberspace

David Bird

Abstract   The dawn of computer-mediated communications (CMC) has been
revolutionary and at the forefront of advancing human collaboration
during this modern Information Age. In our bid to communicate and
share information, our information technology (IT) journey has
transgressed the traditional client-server model and manifested itself in
the cloud. Technological advancements we take for granted today have
blurred the lines of our intellectual property sovereignty within
cyberspace. This chapter discusses my contemporary reflection of the
last 25 years or so of the World Wide Web (WWW) and the Internet that
underpins it – the dominance of this virtual dominion, the future to
behold because of it and the advantages and disadvantages of using it.

Keywords   collaboration; history; IT; Internet; society

Introduction

I have been lucky to have lived through this unique period of mankind's
journey into cyberspace. My participation has been both as an early
adopter of technology and in many cases at the cutting edge of securing
customer systems and networks. Our desire for diversity and
convenience has put intuitive human-machine interfaces on a
progressive journey that spans from the 1980s into the early 21st
century. Communication across the global footprint of the Internet, and
through the use of hypertext transfer protocol (HTTP), has enabled users
to interact more so than at any time before in human history. This

transposition has been captured through productions such as: (i) *War Games* in 1983, (ii) *The Net and Hackers* in 1995, (iii) *We've Got Mail* in 1998, (iv) *Swordfish* in 2001, (v) *Firewall* in 2006, (vi) *The Social Network* in 2010, (vii) *The Fifth Estate* in 2013, (viii) *Halt and Catch Fire* in 2014, (ix) *Steve Jobs and Blackhat* in 2015, and (x) *Mr Robot* in 2015 and 2016. No matter the accuracy of technological fact presented in them, popular culture has helped raise our awareness and acceptance of the online virtual dominion.

In 1995, there were 16 million Internet users and in 2016 there were 7.5 billion Internet consumers (Price Waterhouse Coopers, 2016). It was the rise of smart devices in the new millennium that opened up cyberspace for truly massive adoption – to the point that 50 billion net-enabled devices are expected to be connected by 2020 (Kleiner, 2014). It could be surmised that the psyche of online adoption is intrinsically linked to the widespread and successful adoption of collaborative technologies. My journey in this area has been written with a backdrop of modern computer history in order to provide a view of human collaborative progressiveness through the Internet and WWW and what the future may hold.

The early years

My introduction to the realm of micro-computing began in the early 1980s with the BBC Micro. I also dabbled with the Beginner's All-purpose Symbolic Instruction Code (BASIC) programming language on various International Computers Ltd Personal Computer series machines. Compared to the world of mainframe computing, the emergence of micro-computing was assumed to be a specialist area – almost niche for enthusiasts – at the start of the decade (BBC2, 1982). This movement expanded to many entrepreneurial companies and was coined the "fourth industrial revolution" (Berting et al., 1980). The two Steve's – Jobs and Wozniak – of Apple Computers and Bill Gate's Microsoft were at the forefront of making micro-computing more inclusive to the masses (Channel 4, 1996). Even with slow data rate dial-up communications, niche groups and academia fervently collaborated through the medium of the Bulletin Board Systems (BBS). Unlike their mainframe and minicomputer brethren, micro-computers were standalone clients and this technology quickly changed the world of personal computing.

Until now the computing evolution had been missing a reliable and accessible interconnecting communications path. By 1990, a dramatic event occurred, which I would argue went mostly unnoticed by the majority of the general populace. This event was the simultaneous decommissioning of the US Defense Advanced Research Projects Agency (DARPA) packet-switched ARPANET of the 1960s to form the world's internetwork. Coupled with the European Organisation for Nuclear Research (CERN) servers that hosted the world's first hypertext documents (Berners-Lee, 2000), the Internet and subsequent WWW were spawned from a pedigree of academic collaboration between the United States and Europe that dated back to the early 1980s (Cerf, 1994).

Micro-computing caused an Internet boom around 1993 and personal computers (PCs) were fast becoming a *tour de force* by the mid-1990s. Common Internet Service Providers (ISPs) at the time were CompuServe, America Online (AOL) and Demon Internet who eventually made the Internet more accessible across the globe. I have recollections of the United Kingdom's early adopters using analogue modulator-demodulators (modems) to connect to their early online ISPs using dial-up at around 2400 bits per second (bps) over the Public Switched Telephone Network (PSTN). Due to incessant PSTN lag, the faster 64 kilobits per second (Kbps) Integrated Services Data Network (ISDN) gained some traction in the mid-1990s.

Birth of cyber consciousness

A stoic appreciation of the norms such as telephony, video-on-demand and high-definition television technologies, originally seemed to take conscious primacy over the gradual adoption of CMC (Lyons et al., 1993). However, people soon learned that they could use computer technologies of the day to not only send emails, but to chat on Usenet Newsgroups and search the disparate newsgroups with Gopher, socialize using Internet Relay Chat or through legacy BBSs (BBC, 1995). Personally, I have fond memories of engaging seriously with computing technology during the early 1990s, honing my IT prowess across multiple operating systems, ranging from early Windows to OS/2 and Xenix to SunOS Unix to Linux. But notably the most influential technology, invented as a by-product from CERN (Berners-Lee, 2009) was the vehicle to share and consume hypertext information: the fledgling WWW now known by the retronym Web 1.0.

Into the second half of the 1990s, I was part of the office automation bandwagon. Key tools during this period were termed 'killer apps'. These applications were highly prized and ultimately enabled authors to be more productive. Of course, the original 'killer apps' had already been invented in the 1980s: these were the predecessors of the key productivity proponents that we would recognize in an Office Suite today. In the early 1990s, the notable up and coming killer apps were not only the word publishers, such as FrameMaker, but the advent of the web browsers such as Mosaic (Anderson, 2013) and Netscape Navigator. These browsers centralized the ability to use a graphical user interface for multi-protocol communication across the Internet and fervent browser rivalry fuelled the web browser wars that still continue to this day. Application packages in their various forms facilitated the sharing of data via various mediums: (i) 1.44 MB floppy disks, (ii) email which was becoming a more available means of communication, (iii) uploading files onto websites, or (iv) sharing copy-and-paste text through Instant Messaging (IM). Proprietary Internet-orientated application clients, using proprietary protocols, became the order of the day by the mid-1990s for peer-to-peer and chatroom communications such as AIM, ICQ and MSN.

Unfortunately, at the outset the Internet's founding fathers had used in-secure protocols within what was then a closed inter-networking system. That is, the stack utilizing Transmission Control Protocol (TCP) over Internet Protocol (IP) was not protected. This meant that communications were not private and the issue of objectively inherent trust was discussed in earnest due to perceived network threats (Wallich, 1995). The use of some proprietary protocols could be deemed security-by-obscurity, because without proper *encipherment* they could eventually be cracked. However, between 1994 and 1996, the implementation of HTTP over Secure Socket Layer (SSL) versions 1.0, 2.0, and 3.0 were championed by Netscape and this progressed the foundations for reliable web session encryption. Not only that, oversights during this period, which even resonate today, comprised operating systems' post-installation lockdown that tended to be open by default. I have memories of adding security through third-party applications to bolster the deficiencies of these host operating systems. Additionally, some of these applications were themselves open to controversies through detected vulnerabilities from well-known antivirus vendors, raising the spectre that security applications could be the weak link in PC security. Well-publicized events such as the Morris and the Melissa worm outbreaks, of 1988 and 1999 respectfully, also revealed the far-reaching impact that malicious threats could have on our virtual online dominion.

Teleworking had become a powerful buzz word – following on from the tele-cottage concept introduced in the 80's – providing an opportunity to use technology to decentralize workers from their normal workplaces

(Grootings, 1986). Because home-working was a paradigm shift from an employer's point of view, some arguably thought it would be a fad. But teleworking promised better productivity, contrasted against preconceived ideas of its environmental benefits; these were all pitted against the potential sociological impacts such as psychology and behaviourism (Bradwell and Murray, 1994; Orvice and Ingram, 1995). Conflicting social theories also prevailed. Technological determinism defined technology as a social influencer, whereas structuralism argued that individual choices were driven by social forces. As IT matured, I discovered a third dimension driven by Artificial Intelligence (AI); in this arena, I used high-level programming fuzzy-logic rules and virtual neural networks to investigate the functions of machine-learning and knowledge-based systems.

By 1997 the globalisation of the Internet, and the WWW overlaid on top of it, took a new turn through the embryonic e-economy (Hartman, 2000). Bill Gates (1995) foresaw that the PCs which influenced the first digital revolution would be the arbiters of the second. I can remember the Press becoming completely infatuated with the dawn of full Internet commercialization – calling it the digital superhighway. I would argue that Gates was partially correct, but he had not factored in the level of influence that portable devices and mobile phones would have in his vision of the future. An example of the former is epitomized by the Psion series of portable digital assistants (PDAs) that trail-blazed palmtop computing. Moreover, technologies such as Nokia company's adoption of the 2.5G Global System for Mobile Communications (GSM) web-compatible Wireless Application Protocol (WAP) was a huge influencing factor for mobility into the new millennium. Personally, I traded in my Matrix-style Nokia 8110 mobile for the 7110, complete with the first mobile WAP browser so I could interact in a limited capacity with the WWW on the hoof. From my perspective, the two things that constrained the uptake of IT was social acceptance and the lethargic data rates at the time shown in Figure 1 on page 258.

This was particularly poignant because concepts, such as 'the digital nervous system' had been aired by Bill Gates (1999) in order to provide the right information to the right people at the right time.

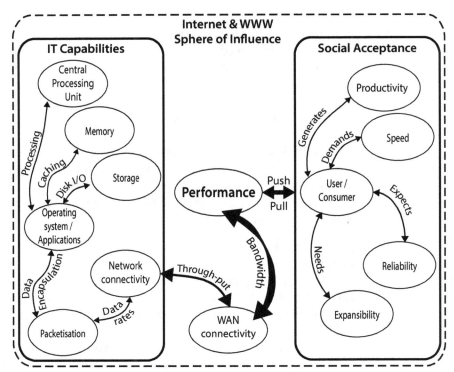

Figure 1: Collaborative causation

Into the new millennium

As our global community had only just started our evolutionary net-centric journey, we had to work together for the common good, to remediate critical systems affected by the potentially catastrophic Millennium Bug (Jones, 2014). By ensuring the survival of the modern computer age, major communications developments flourished into the broadband era. Known as the Asynchronous Digital Subscriber Line (ADSL), this technology was made commercially available in 2000, providing around 2 Mbps. The complementary 802.11b was released in 1999 and manufacturers combined wireless with their ADSL router-modems. Wireless Fidelity (WiFi) naturally soon took hold at the start of the century. As a result, traditional ISPs, who operated as dial-up portals onto the Internet, were surpassed by telecommunications companies (telcos) that provided network gateways connecting directly to the Internet wide area network (WAN).

But what surpassed all previous advancements from a personal perspective was the fusion of PDA-type computing and the mobile phone, creating the first smartphone form-factor. Subsequently, mobility interests reconciled me to adopt various brands of Microsoft Pocket PC PDAs during the early millennial decade, progressing into the Windows Mobile generation of devices between 2002 and 2006. A key accelerant was WAP browsing, which became richer and quicker as mobile data networks developed and the stack progressed. WAP was eventually surpassed with the release of 3G commercial cellular in 2004 that provided speeds up to 384Kbps. Undoubtedly 3G was the catalyst for even wider adoption of Internet-connected devices.

Then in 2007, Apple's iPhone entered the scene. This device was heralded as a game changer because it redefined the smartphone. This was partly because Apple rendered full Hypertext Mark-up Language (HTML), upon which we had already become accustomed since the 1990s through desktop web browsers. The original model of the iPhone OS 1.0 closed operating system was the use of browser-presented web-served applications, complementing the phone's organic apps. It was an approach that followed the principle of Google's 2006 incarnation of web-served software-as-a-service (SaaS) productivity applications. However, Apple soon realized that consumers craved host-installed apps, which won out causing Apple to release a software development kit (SDK) in 2008 and app downloads from the App Store for iPhone OS 2.0. Coincidentally, Google's Android open operating system was unveiled in the same year as the iPhone 1.0 and swiftly challenged Apple's product. By 2008, Android was not only deployable onto multiple vendor smartphones, through the Open Handset Alliance initiative, but also encouraged app downloads from the Google Store. Today both Android-centric manufacturers and Apple fight for dominance in the smart-device market that incorporates not only smartphones, but also tablets.

A key influencing factor for later web back-end services was the introduction of Utility Computing around 2001. This provided the springboard for reliable hypervisor-based virtualisation technologies that can be either 'bare metal', such as VMWare, or host-based like Xen and the subsequent Kernel Virtual Machine. The subsequent Service Orientated Architecture concept drove Infrastructure-as-a-Service models into clouds on the Web, followed by SaaS application hosting. These formed the lynchpin of cloud services and the onset of online capabilities such as Amazon Web Services.

Increased offerings and user demand broke down atypical barriers and awoke our society to the benefits of not only using online-enabled communications but also the advantages of the online-hosted medium.

Dubbed Web 2.0, interactivity between users through the power of the Web became the *nouveau chic* from 2004. I would argue that, based on user adoption, web-enabled social media has now closed the gender gap – through online platforms such as LinkedIn, Facebook, Twitter and also Instagram. Cyberspace has been the catalyst for a greater use of collaboration tools that include Voice over IP (VoIP) and web-enabled video conferencing. This evolved collaborative capability benefits both collocated and virtual teams alike, crossing time, space and cultural boundaries (The Open University, 2016). It could be said that interactions between scientific, technological, economic, social and cultural forces have all had their part to play (King, 1984).

Second post-millennium decade

Strides in telecommunications over the past decade have undoubtedly provided greater bandwidth, such as advances in ADSL increasing through-put from 2 Mpbs to nigh on 50 Mpbs for fibre-based very-high-data-rate digital subscriber line technologies (VDSL 2.0). 4G cellular has also been launched within the past three or four years and provides a quantifiable leap in data rates over 3G up to 1 Gigabit per second (Gbps). From my personal observation, society is now transforming, partly through the upsurge of smart-devices and what could be called the 'smartphone wars'. Research in Motion, once known for secure enterprise mobile phones, and the popular IM Blackberry Messenger, arguably failed to keep up with the competition, losing market share, and succumbed to a lack of functionality and available apps. Android smart-device derivatives presently hold the greatest market share of smart-devices but a huge number are inherently proven to be vulnerable (McGoogan, 2016). Apple, also had their spate of weaknesses over the years, but to evolve they reliably provide the most widely available operating system updates to remediate vulnerabilities.

Interestingly, smart-device facilitated social media apps provided a platform for ordinary people – spanning many North African and Middle-Eastern countries – to stand up for their rights during the Arab Spring of 2011. Unfortunately, social media was also used in the UK to enable looters to organize themselves during the summer riots of the same year. These events reinforced the power and diversity of social media, which appears to follow Stanley Milgram's Small World principle (1967) of six degrees of separation between people in a social network. These circumstances have brought about techniques such as Social Media Intelligence to the fore; a capability that can be used to analyze social networks, and liaisons, through social media provider application programming interfaces (APIs) or simply through online searches against open user profiles. My view is that now that this genie is out of the bottle, it will be extremely hard to put it away again.

Free online services exist in many guises and social media is just one example making us, the consumer, 'their product'. We are effectively treated like commodities and their intention is to harvest our data usage, our emails, our web searches, and maybe even our online storage so that companies can target us with adverts. Immense amounts of data are gathered about our personal preferences through 'big data' processing to the point where nothing is private anymore (Zlawdon, 2015). After the infamous Snowden revelations of 2013, Internet privacy has been injected into the public's consciousness and users are becoming more security savvy about applying secure settings to their Facebook, LinkedIn and Twitter accounts (Bird, 2013a). As an outcome of the legacy ethos regarding Internet information freedom and openness, it is recognized that there should not be an expectation of privacy unless we are paying for the requisite online service (Paterson, 2016). We should, therefore, be careful about what we are willing to give up online for free (Creese, 2015). A recent survey by LaunchTech has indicated that millennials, with their blasé online-centric approach to life, are more likely to become insider threat actors of the future (Lawton, 2016).

Various forms of encompassing security layers can provide some form of protection to prevent data from being unnecessarily exposed or stolen. Virtual private networks (VPNs) are now popular but actually first appeared to the general population back to the mid-1990s. Cyphers have grown in popularity over time, culminating in the wider use of the Advanced Encryption Standard (AES) introduced back in 2001. Following the current trend of public concern over Internet privacy, the advertising industry has been forced to recognize that less intrusive advertising, less resource intensive and more respectful ads are the way forward for the Web (Ryan, 2016). As a result of email privacy concerns, ProtonMail, a spin off from CERN, has created a webmail service to protect our emails online by using multiple layers of encryption. That said, recent newsworthy items like the infamous Heartbleed debacle, which is based around vulnerable variants of SSL, have resulted in visibly more modern data-in-transit encryption methods being used, such as Transport Layer Security (TLS) version 1.2 crypto algorithm (Man, 2015).

Web 2.0 has been about web content richness. The Semantic Web, which is in its infancy, is based on the use of interrelated meta-data – enabling humans and machines – to interpret the structured content destined for Web 3.0 (Carr, 2016). This next web idea, promoted by Sir Tim Berners-Lee in 2007, takes the Web down the road of making linked data more searchable, and useable, potentially making it understandable to computers through the use of data-class ontologies. Web services have become an important feature to enable endpoints to interface with data centre-based system back-ends using exposed online APIs; this approach was initially realized through Simple Object Access Protocol (SOAP) over transports such as HTTP, Extensible Mark-up Language messages and now through a more modern replacement called the lighter weight JavaScript Object Notation (JSON) format. These mechanisms are used for data distribution and backend application parsing of both isolated and public data depending on the use-case. Already cloud providers have jumped onto the 'big data' bandwagon, promoting their services so that customers can conduct the collection, storage, processing, and analysis of data at scale. The emergence of the Internet of Things (IoT) over the virtual horizon brings another new dimension, building upon established web-enabled data exchange methods. This journey is shown in Figure 2.

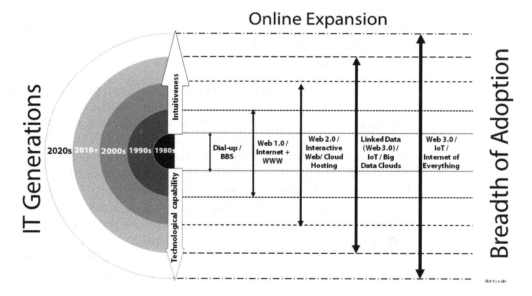

Figure 2: Online expansivity

Transformation to our online-centric approach is not all about the platforms but the applications that run on them and our interactivity with online services. This progression has changed the threat landscape and the level of potential hacker accessibility to our information. Inevitably, code development and distribution packages (distros) in the cloud are now becoming the norm; however, recent nefarious activities continue to highlight the Black Hat hacker community's appetite to subvert shared online source-code repositories (Clark, 2015) on the searchable WWW – known as the Surface Web. But just underneath there is what is known as the Deep Web, where collaboration can be performed between undeclared closed groups conducting generally non-illicit activities. However, the next level down is the Dark Web where notorious, nefarious and illegal enterprise takes place on the fringes of the Internet. The Dark Net is accessed off-grid by using anonymising obfuscation techniques such as The Onion Router (Ingram, 2016); here malign services such as selling narcotics on the Silk Road site or transactions of skimmed credit card details are atypical of the Dark Web (Millman, 2016). The use of such web sites on the fringes makes national borders porous and sites hard to tear down by law enforcement (Dean, 2014). However, as a consequence of WikiLeaks revelations, advocates are now promoting techniques associated with the Deep Web to achieve general anonymity, making users less attributable online (Winter, 2015).

Our use of online services places a certain amount of responsibly upon service providers who should consider the following: (i) more use of trust-based account credentials, and (ii) a level of trust that providers' web-enabled services are not themselves vulnerable (Anderson, 2016). This is particularly important since the inception of Google online productivity apps in 2006 and Microsoft's more recent evolution of Office into the Office 365 'always cloud-connected' variant. But in order to be a part of the future, customers who host applications online must themselves ensure they are not unnecessarily vulnerable in the cloud and are responsible enough to deploy proportionate security controls to protect their cloud instances (Westervelt, 2013). Doing the opposite just increases the attack surface and the spectre of external malicious hacks.

Projections for the future

Today Unified Communications has come to the fore as an enabler from technological advancements, mobility and ease of use. Today, the once-dominant functionality-rich PCs are in decline. So, will the ensuing smart-device trend, and Google-type slim-clients, continue the incessant gravitas provided by their intuitive ease-of-use over traditional thick-clients? Interestingly, Apple appears to be erring towards more smart-device innovation. So, what does Microsoft have up its sleeve for its successful hybrid Surface product line after 2025 when their cross-platform Windows 10 operating system is purportedly discontinued? In

IT, five years is a long time and in my mind convenience still wins the day – especially now when smart-devices are nigh on becoming consumable items and it is feasible to perhaps make the transition towards credible mixed-reality devices (Tung, 2017).

As part of this upsurge in collaborative interactivity – spanning human and technological interaction – a diversity of communications delivery methods has been tested in the guise of wearables. By late 2016, innovative wearables, such as smartwatches and Google Glass, have thus far have been generally dependent on personal area wireless tethers to parent smartphones. That said, more recently such devices appear to be stagnating except for compatible fitness monitoring derivations (Keach, 2016). Although, previously attempted with iPhone 1.0, and reinvigorated through web-centric Chromebook slim-clients, perhaps web-integrated devices will come back to the fore with the advent of cloud-centric Mobile-backend-as-a-Service technologies. Facebook's projection of the future in the social media space is that greater video immersion will become the medium of choice for human interactivity, followed by Virtual Reality (VR). Does this mean we are going to wear VR goggles as the next phase to interact with applications such as Pokemon Go? Probably not; the Google Glass experiment has demonstrated that this may not be the case for now but the next generation of smart-devices are anticipated to become more dominant through their emergent properties and ubiquity (Checkland, 2007). It has been proven that the social conditions of different social systems influence the ways in which new technologies are embraced (Grootings,1986). As a case in point, common technology is espoused in different ways based on socio-diversity and sociability shown in Figure 3 on page 265.

Unlike Web 2.0, the 'web of data' or the Semantic Web will enable data to become more interpretive, interoperable, relationship-based and interrelated as 'linked data'. Current technology was originally geared to search, surf and sift through millions of indexed web pages; today, however, there are billions, so existing approaches will not scale upwards of billions of billions and even trillions of web pages in the future (Spivack, 2015). The Semantic Web has been encapsulated by the Web Ontology Language standard coupled with the Resource Description Framework, forming an extensible knowledge representation data model for open data sharing (Simperl, 2016). The original Web 3.0 concept drives towards a data superset, which is expected to follow the open model for context and data sharing that already exists – but in a more intelligent way using AI. As a consequence, I believe there should be caution as to what kind of data is exposed through such approaches. In

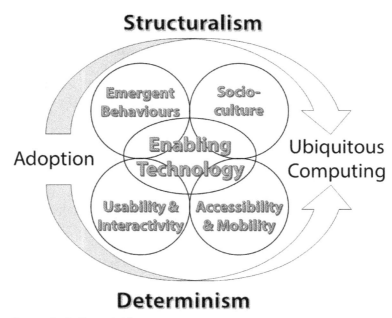

Figure 3: Influential forces

fact, there is actually a schism about agreeing the way ahead for the reframing of the Web in such an open manner to make data useful and publicly available. In my view, a mechanism would need to be in place to avoid the use of misrepresented data so that incorrect or misleading data cannot be interlinked. In addition, the use of isolated datasets will still have a place under certain circumstances, as it has to date, for the processing and storing of personal data, non-public governmental data and niche scientific data.

The concept of using the WWW as a gigantic collaborative knowledge base was conceived during the mid-1990s and a combination of the Semantic Web and heterogeneous AI web-centric 'big data' crunching capabilities could arguably be another progressive leap for mankind. Already Google, Amazon Web Services and Microsoft Azure are posturing for AI dominance, making it potentially the next battleground (Schwartz, 2016). Beyond present algorithms that we take for granted, more advanced AI implementations by Microsoft and Google are purported to be using either field-programmable gate arrays or non-programmable application-specific integrated circuits; these in conjunction with interrelated and interoperable datasets, could potentially start a new digital information evolution, where AI data reasoning could be used for the good of everyone (Cooper, 2016). Currently, cloud is portrayed as the future, ranging from application containerisation to scalable capabilities for 'big data' ingestion, pre-processing, queuing, transposition, post-processing and storage using ephemeral or persistent means. That said, fail safes would need to be introduced pertaining to machine-generated data and machine-interpreted knowledge. Erroneously altered or misinterpreted data could cause unexpected decisions or unintended outcomes from 'big data' analytics: integrity is key (Palmer and Gent, 2016).

Although machine-learning has already been integrated for cyber defence, AI is being touted as a prominent capability for the ensuing IoT security vigilance conundrum (Golan, 2016). Conversely, there are predictions that AI will be used for cybercrime, whether it be conning people through chat-bot techniques or potentially AI systems themselves becoming cybercriminal actors in their own right (Mitchell, 2016). Comparably, unlike AI developed to conversationally emulate humans as part of Turing's Imitation Game, Microsoft's Tay was developed to learn from human online interaction. Unfortunately, Tay's machine learning predominately focused on negativity that has caused concern over our own humanity (Horton, 2016). Contrastingly, Google's Deepmind AI system uses 'friendly AI' artificial neural reasoning, with extended memory, to lock away data nuggets for recall later. Deepmind's collaborative governance oversight provides an approach for this technology to benefit mankind (Merriman, 2016). Other applications of AI include mathematical reasoning using neural networks to crack recently developed homomorphic cryptographic schemes; homomorphic cryptographic techniques are purported to allow data to be processed without actually decrypting it: ironically, AI provides a mechanism to break some schemes even before they have been used in anger (Bogos et al., 2016).

In a similar vein, Quantum Computing, which has a pedigree that goes back decades, has entered into the fray as a 'frenemy' of cryptography. Quantum is a serious contender to crack public key cryptography (PKC). PKC is in itself the main encryption mechanism used online and the ability to decisively break it in the future will certainly prove divisive. Contrariwise, it was originally thought that quantum cryptography could only be useful to encrypt data over short distances. However, the Chinese have recently launched a satellite to test the viability of using quantum-generated keys to encrypt transmissions over long distances, thereby rendering communications immune to attack (Macleod, 2016). Interestingly, Google has risen to the challenge of conducting research into quantum cryptographic technology, as a gatekeeper to safeguard robust trust credentials for the Internet (Templeton, 2016).

The IoT has already arrived and is categorized by four types of data types: (i) dumb sensor, (ii) near real-time monitoring, (iii) smart two-way communicative devices, and (iv) real-time processing by AI capable intelligence. A water meter would be an example of the former and an autonomous self-driving vehicle would be at the latter end of the IoT capability spectrum. IoT mechanisms are being developed such as: (i) Extensible Messaging and Presence Protocol originally designed or human collaboration but is now being transformed for human-to-machine and machine-to-machine (M2M) exchanges, (ii) Advanced Message Queuing Protocol for M2M broker-centric publisher and subscriber distribution methods, and (iii) other Lightweight M2M derivatives for sensor or device inter-communications. Unfortunately, the IoT race to market has, to date, arguably precluded adequate security-by-design measures, which is ironic when IoT is driven by the human technological desire for expansive 'data insights'. Recent security concerns over employing flawed encryption implementations in IoT devices has already caused consternation; where, for example, two million devices were found to be remotely vulnerable (Newman, 2016). Other risks also prevail such as M2M communication techniques and online service provider terms and conditions that may erode our data sovereignty and privacy when used to create, store and share our digital content. This is particularly poignant as we enter the Web 3.0 era, where computers could exchange 'linked data' for the IoT generation (Matusky, 2015). The use of 'data insights' could also cause massive reverberations, some of which could be about unexpected insights into our personal habits (Tanner, 2009). To meet the increasing demands and anticipated increase in IoT devices by 2020, the UK will need to evolve the current cellular technologies to a post-4G generation for advanced long-term evolution.

Recent research indicates that there are presently two and a half billion smartphones globally and this is anticipated to reach six billion by 2020 (Golan, 2016). It is my belief that new cellular communication technologies such as 5G will be the arbiter to bring Cisco's prediction of an Internet of Everything (IoE) closer to fruition. 5G is expected to achieve data rates of more than 10Gbps and has been heralded as the credible dynamic multi-path low-latency network concept for many IoT devices. 5G potentially introduces concepts such as Device-to-Device communications of which M2M communications and even Vehicle-to-Vehicle communications are spin-off derivatives. In the 5G age new methods of identity and key management by mobile service providers would be a necessity to achieve 5G aspirations. Trusted device boot sequence, using hardware anchors and integrity verification to the 5G

Radio Access Network (RAN), would ensure that any keys associated with the 5G network and the user device itself are legitimate. Traceable trust would be prudent to avoid basestation (BS) spoofing, or hiding, and data snarfing (grabbing a large document file, content or data and using it without the owner's permission) when users move between 5G cells. Concepts such as separating the user data plane from the system control plane in Software Defined Networking across the evolved core enables optimization for both handsets and the mobile networks themselves. Network functions virtualisation, itself comprising interfacing virtual network functions, will enable abstraction and micro-segmentation approaches from the interfacing RAN BS to the next generation core containing virtualized middle-box applications. Obfuscation methods, similar to existing networks, to protect consumer identifiers from outward disclosure to the public Internet are still relevant in principle for the 5G-evolved mobile core. Onward-routed IP traffic to the Internet from the 5G telco mobile networks will still need to pass through security appliances at their edge gateways (Bird, 2015).

Conclusion

My historical 'back to the future' account just scratches the surface of our infatuation with cyber. Building upon the foundations laid down in the 1990s, online media broke new ground against established communications methods. Subsequent technological advances enabled collaboration and brought us a new dimension for productivity and innovation. Now IT capability has transitioned from email at one end of the spectrum to cloud-deployed server-less web applications at the other. All in all, I think the key proponent for the successful foothold and uptake of IT by our civilisation has been the computer processing articulated by Moore's Law combined with the speed of data access, reliability, compatibility and bandwidth (Arthur, 1994).

Although PCs had saturated the market, it was really the onset of the smartphone that was the major key enabler (Miller et al., 2016) for greater engagement and utilization of the WWW beyond just consuming content. In my opinion, technology specifics that were important in the 1990s to quantify computing power have been surpassed by functionality that enables users to engage with content online. My view is that Web 2.0 has been the instrument to reap the benefits in the early 21st century and springboard teleworking into a more credible working *modus operandi*. As a global civilisation, we can now either work as collocated teams or separately at a distance because technology has the ability to collapse time and space (Millard, 2016). A vital factor in an information economy is the fact that information can be shared and communicated to the right recipients – whether human or machine – at the right time. A key success factor of such technology is not only the collaborative aspects but also trust in such technology; however, it can have negative side effects by creating Internet addicts, especially in teenagers, who can become hooked on their fix of daily online accesses (Goodwin, 2015). The journey we have been on is a fusion of technological advancement and our interactivity with the online world – a virtual dominion that is now directly controlling physical things in the real world through operational and IoT technology.

Today the Internet is used as a huge and unrivalled collaboration network. With job losses from the stagnating PC market in 2016 (BBC, 2016), it could be said that smart-device functionality and portability now provides a more productive user experience online. I am conscious through my own personal preference that I tend to use smart-devices more readily, another statistic of this trend for our Internet-connected accessibility. Weighted evidence from my personal studies in the 1990s suggested that technology adoption by women tended to be based on technological usefulness rather than the male view of embracing technology as if it were a toy. Social media has transgressed the divide beyond technology enthusiasts and has become a liberating effect especially for women (Cadigan, 1994) to the point where addictiveness is raising concerns over mental health (Burns, 2016). Social media has been an evolving format to the point where the use of certain social media variants like Twitter has morphed from a sociability user-base towards a knowledge sharing and scientific-slanted consumerism (Burgess, 2015). These days cyber is so ingrained that it is now extremely hard for employers to prevent workers from getting their daily sociability fix during working hours.

Web 1.0 was about information access to published benign data and Web 2.0 followed where richer interactivity and participative collaboration is readily available. It has been identified through research that social media adoption differs in different areas of the world (UCL, 2016). The importance of social media has now spawned academic socio-technologist disciplines such as Web Science. Usefulness and the application of social media is being demonstrated through academic studies, based around social media platform affordances and its features (WUT, 2016). Unmistakably, though, the WWW is also used by radicalized fundamentalists to wage a propaganda and psychological war and to recruit terrorists (Neumann, 2016).

Unfortunately, a fundamental problem has been the original design of the Internet, from an open systems perspective, making our virtual world susceptible (Bird, 2013b), especially when the Internet was not designed to be the central nervous system of modern life (Lucas, 2016a). It is a myth that it is possible to have a completely secure system, but we should strive to be proportionately secure against the established threat landscape, and update our software so that we can plug undiscovered vulnerabilities that may materialize over a protracted period of time. Now corporate and personal VPN services have become a factor of life in cyberspace to preserve some semblance of personal online privacy. In 2016, social media platforms began adopting end-to-end encryption for private messaging. Today identity assurance is weak and there are still issues around identity and user trust verification where web-orientated two-factor authentication (2FA) adoption has been slow. Trust online today has been described to be akin to a 'Masked Ball' analogy (Lucas, 2016b); where peoples' identities may or may not be truly verifiable and they may not be who they said they are or who you think they should be. In the future, we need to be able to implement technology add-ons not only to ensure privacy but also trusted integrity.

Our desire for interconnectivity has also brought about maliciousness that used our legacy collaboration tools as a method of propagation, such as the ILOVEYOU virus in 2000, Code Red in 2001 and the Conflicker worm in 2009. The recent spate of high-profile hacker attacks against web services go to reinforce the lengths that hackers will go to acquire information. The hacker impact upon an organisation's reputation are quantified through the following examples: Twitter losses (Shu, 2016) and LinkedIn breaches of 2012, one billion Yahoo accounts compromised in 2013 (Thielman, 2016), the TalkTalk compromise and Ashley Madison attacks in 2015 (Bird and Channel 4, 2016), and the recent Panama Papers information leaks (Harding, 2016).

Whether we like it or not, technological change will continue to occur and it is those who adapt to change who will make the most of new opportunities that change offers (Shirley, 1986). As we progress we need to learn lessons from past malicious intent or their resultant outcomes in order to safeguard our progressive Internet future. Now the Web 3.0 era has dawned, data sharing will only propagate through the intertwined journey of technological advancement and cyber adoption reflected in Figure 4 on page 271.

We must be cognisant of the fact that the online virtual revolution is the biggest social experiment in human history (Krotoski, 2009). Coupled

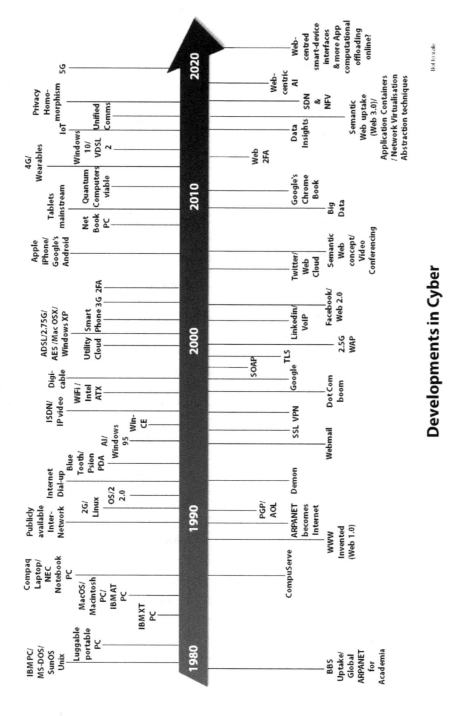

Figure 4: Collaborative technology timeline

with behavioural change, trust, privacy and security protection of user and M2M activities, an increase in excess of 50 billion Internet-connected devices is inevitable by the 2020s; connections not only between humans, but between humans and machines, and among machines for AI-based computation. In a complementary way, the next revolution may not be the hype around Web 4.0 but actually the advancement of AI and its wider collaborative use. This will be facilitated by data exchanges and distributed communications across cyberspace affecting an IoE epoch.

References

Anderson, T. (2013). *Mosaic turns 20: Let's fire up the old girl and show her the web today*. Available at: http://www.theregister.co.uk/2013/04/26/mosaic_20_anniversary/ [Accessed 14 Feb 2017].

Anderson, T., (2016). *Linux Mint hacked: Malware-infected ISOs linked from official site*. Available at: http://www.theregister.co.uk/2016/02/21/linux_mint_hacked_malwareinfected_isos_linked_from_official_site/ [Accessed 14 Feb 2017].

Arthur, C. (1994). *Visions of Netropolis*. New York, USA: New Scientist Supplement. pp1–31.

BBC. (1982). The Computer Programme Episode 2. BBC2. Television.

BBC, (1995). The Net Episode 5, BBC Television.

BBC business. (2016). *HP cuts up to 4,000 jobs worldwide*. Available at: https://www.google.co.uk/amp/www.bbc.co.uk/news/amp/37651831 [Accessed 14 Feb 2017].

Berners-Lee, T. (2000). *Weaving the Web*. London, UK: Texere Publishing Ltd, p2.

Berners-Lee, T. (2009). *The Next Web*. Available at: https://www.ted.com/talks/tim_berners_lee_on_the_next_web [Accessed 14 Feb 2017].

Berting, J., Mills, S., and Wintersberger, H. (1980). The Socio-economic Impact of Microelectronics, Pergamon, Oxford. In Open University, *Information Technology: People and Issues Block 1C Part 2*. (p11). Open University Press.

Bird, D. (2013a). *Open or Closed*. Swindon, UK: BCS ITNow magazine.

Bird, D. (2013b). *Underlying Dangers*, Swindon, UK: BCS ITNow Magazine.

Bird, D. (2015). *5G: the need for speed*. Swindon, UK: BCS ITNow Magazine.

Bird, D. (2016). The Age of the SoC. Swindon, UK: *BCS Digital Leaders Magazine*.

Bogos, S., Gaspoz, J., and Vaudenay, S. (2016). *Cryptanalysis of a Homomorphic Encryption Scheme*, Lausanne, Switzerland: EPFL.

Bradwell, D. (1994). *Starting working from your home*, Ziff Davis UK: Personal Computer Magazine, pp87–91.

Burgess, J. (2015). *Twitter (probably) isn't dying but is it becoming less sociable*. Available at: https://medium.com/dmrc-at-large/twitter-probably-isn-t-dying-but-is-it-becoming-less-sociable-d768a9968982#.3ow061rsa [Accessed 14 Feb 2017].

Burns, J. (2016). Girls becoming increasing unhappy – study. BBC News.

Cadigan, P. (1994). The Net. BBC Television.

Carr, D. Prof. (2016). Web Science, University of Southampton. FutureLearn Online.

Cerf, V. (1994). *Networks*. New York, USA: Scientific American Magazine.

Channel 4, (1996). Triumph of the Nerds Episode 2. Channel 4 Television.

Channel 4. (2016). Sex, lies and Cyber Attacks. All 4, Channel 4 Television.

Checkland, P. (2007). *Thinking through systems thinking*. London, UK: Routledge. pp49–51.

Clark, J. (2015). *GitHub code repository rocked by 'very large DDoS' attack*. Available at: http://www.theregister.co.uk/2013/08/15/github_ddos/ [Accessed 14 Feb 2017].

Cooper, M. (2016). *Do you think technology is effecting our society, children and relationships?* BCS Online.

Creese, S. Prof. (2015). Hacking the Internet of Things, The Times Science Festival.

Dean, J. (2014). *Business as usual on 'Dark Web' one week after sting*. London, UK: The Times.

Gates, B. (1995). *The Road Ahead*, London, UK: Viking. pxi.

Gates, B. (1999). *Business @ the speed of thought*. London, UK: Penguin. ppxiii–xxii.

Gent, E. (2016). *Big Data's Dark Side*. IET London, UK: E&T Magazine.

Golan, Y. (2016). *How AI will transform cyber security*. Available at: http://venturebeat.com/2016/11/22/how-ai-will-transform-cybersecurity/ [Accessed 14 Feb 2017].

Goodwin, D. (2015). *My nephew smashed up his parents' bedroom because they turned off the Internet*. London, UK: Daily Mail Newspaper.

Grootings, P. (1986). Technology and Work: East-West Comparison, Groom Helm, London. In Information Technology: People and Issus Block 1C Part 2 (p45). Open University Press.

Harding, L. (2016). *What you need to know about the panama papers*. Available at: https://www.theguardian.com/news/2016/apr/03/what-you-need-to-know-about-the-panama-papers. [Accessed 14 Feb 2017].

Hartman, A. and Sifonis, J. (2000). *net ready*. New York: USA: McGraw-Hill. xx.

Horton, H. (2016). *Microsoft 'teen girl' AI turns into a Hitler loving sex robot within 24 hours*. Available at: http://www.telegraph.co.uk/technology/2016/03/24/microsofts-teen-girl-ai-turns-into-a-hitler-loving-sex-robot-wit/ [Accessed 14 Feb 2017].

Ingham, C., (1995). *Working well at home*. London, UK: Thorsons Business. p2.

Ingram MBE, P. (2016). *Dark Web – Ignore this Security Space at your Peril*. Available at: http://www.securitynewsdesk.com/dark-web-ignore-security-space-peril/ [Accessed 14 Feb 2017].

Jones, J. (2014). *How the UK coped with the millennium bug 15 years ago*. Available at: http://www.bbc.co.uk/news/magazine-30576670 [Accessed 14 Feb 2017].

Keach, S. (2016). *Fitbit is buying Pebble – read the official explanation here*. Available at: http://www.trustedreviews.com/news/fitbit-buys-pebble-acquisition-confirmed-official-statement [Accessed 14 Feb 2017].

King, A. (1984). *The coming Information Society*. London, UK: The British Library. p1.

Kleiner, T. (2014). *5G Research in Horizon 2020*. Europa Publishing.

Krotoski, A. Dr. (2009). Virtual Revolution, BBC Television.

Lawton, S. (2016). *Millennials changing the face of cyber-security*. Available at: http://www.scmagazineuk.com/millennials-changing-the-face-of-cyber-security/article/568862 [Accessed 14 Feb 2017].

Lucas, E. (2016a). Cyber Security vs Personal Privacy, The Times Science Festival.

Lucas, E. (2016b). *Cyberphobia*, Kindle Edition, London, UK: Bloomsberry Publishing. p1256.

Lyons, M., Cochrane, P., and Fisher, K. (1993). Teleworking in the 21st Century. *Computing and Control Engineering Journal*, 46–49.

Macleod, M. (2016). Beijing launches 'hack-proof' satellite. London, UK: The Times Newspaper.

Man, J. (2015) *PCI-SSC Announces the End of SSL Usage for the Payment Card Industry*. Available at: https://www.tenable.com/blog/pci-ssc-announces-the-end-of-ssl-usage-for-the-payment-card-industry [Accessed 14 Feb 2017].

Matusky, R. (2015). *Web 2.0 vs Web 3.0 – What Really is the Difference*. Available at: http://randymatusky.com/2015/04/03/web-2-0-vs-web-3-0-what-really-is-the-difference/ [Accessed 14 Feb 2017].

McGoogan, C. (2016). *Bug leaves 900 million Android Devices vulnerable to attack*. Available at: http://www.telegraph.co.uk/technology/2016/08/08/bug-leaves-900-million-android-devices-vulnerable-to-attack/ [Accessed 14 Feb 2017].

Merriman, C. (2016). *Googles AI Deepmind can now travel the London Underground*. Available at: http://www.v3.co.uk/v3-uk/news/2474074/googles-ai-deepmind-can-now-travel-the-london-underground [Accessed 14 Feb 2017].

Milgram, S. (1967). The Small-World Problem. *Psychology Today*, 1(1), 61–67.

Millard, D. Dr (2016). The Power of Social Media, University of Southampton. FutureLearn Online.

Miller, D., Cost E., Haynes, N., McDonald, T., Nicolescu, R., Sinanan, J., Spyer, J., Venkatraman, S., and Wang, X. (2016). *How the World Changed Social Media*. London, UK: UCL Press, p27.

Millman, R. (2016). *Nearly 6000 online stores hit by hackers*, Available at: http://www.scmagazineuk.com/nearly-6000-online-stores-hit-by-hackers/article/548457/ [Accessed 14 Feb 2017].

Mitchell, R. (2016). *AI predicted to commit more cyber crime than people by 2040*. Available at: http://www.allaboutcircuits.com/news/ai-predicted-to-commit-more-cyber-crime-than-people-by-2040/ [Accessed 14 Feb 2017].

Murray, I. (1994). *Get into the hiring line*, London, UK: Daily Express Newspaper. p40.

Neumann, P. (2016). *Radicalized*. London, UK: I.B. Taurus & Co Ltd. p86.

Newman, L. (2016). *AKAMAI FINDS LONGTIME SECURITY FLAW IN 2 MILLION DEVICES*. Available at: https://www.wired.com/2016/10/akamai-finds-longtime-security-flaw-2-million-devices/amp/ [Accessed 14 Feb 2017].

Orvice, V. (1995). *Farewell, family life – the work shift*, London, UK: Daily Mail Newspaper. p21.

Palmer, D. (2016). *Beware the Midas touch: How to stop AI ruining the world*. Available at: http://www.zdnet.com/article/beware-the-midas-touch-how-to-stop-artificial-intelligence-ruining-the-world/?ftag=TRE3e6936e&bhid=25938706913849853718443426927355 [Accessed 14 Feb 2017].

Paterson, K. Dr. (2016). Cyber Security vs Personal Privacy, The Times Science Festival.

Price Waterhouse Coopers. (2016). Operational Adequacy for the General Data Protection Legislation, Cyber UK Event held in Liverpool.

Queensland University of Technology, (2016), Social Media Analytics. FutureLearn Online.

Ryan, J. (2016). Click, BBC Television.

Schwartz, J. (2016). *Cloud leaders diverge on AI approach*. Available at: https://awsinsider.net/articles/2016/10/10/cloud-leaders-diverge-on-ai.aspx?m=1 [Accessed 14 Feb 2017].

Shirley, S. (1986). If it wasn't for people in between, Computing. In Information Technology: People and Issues Block 1C Part 2 (p14). Open University Press.

Shu, C. (2016). *Passwords for 32M Twitter accounts may have been hacked and leaked*, Available at: https://techcrunch.com/2016/06/08/twitter-hack/ [Accessed 14 Feb 2017].

Simperl, E. Prof. (2016). Linked Data and the Semantic Web, University of Southampton. FutureLearn Online.

Spivack, O. (2015). *Web 3.0*, Available at: https://flatworldbusiness.wordpress.com/flat-education/previously/web-1-0-vs-web-2-0-vs-web-3-0-a-bird-eye-on-the-definition/ [Accessed 14 Feb 2017].

Tanner, B. (2009). Click, BBC Television.

Templeton, G. (2016). *Google is testing a Chrome that adds post-quantum encryption*. http://www.extremetech.com/extreme/231520-google-is-testing-a-chrome-that-adds-post-quantum-encryption [Accessed 14 Feb 2017].

The Open University. (2016). *Themes and theories for working in virtual project teams*, Milton Keynes, UK: The Open University Press.

Thielman, S. (2016). *Yahoo hack: 1bn compromised by biggest data breach in history*. Available at: https://www.theguardian.com/technology/2016/dec/14/yahoo-hack-security-of-one-billion-accounts-breached [Accessed 14 Feb 2017].

Tung, L. (2017). *Microsoft: The smartphone is dead. You just don't know it*, www.zdnet.com/article/microsoft-the-smartphone-is-dead-you-just-dont-know-it/ [Accessed 10 May 2017]

UCL. (2016). Why We Post: the Anthropology of Social Media. University College, London. FutureLearn Online.

Wallich, P. (1995). *Wire Pirates*, New York, USA: Scientific American Magazine.

Westervelt, R. (2013). *Amazon S3 users exposing sensitive data, study finds*. Available at: http://m.crn.com/news/security/240151857/amazon-s3-users-exposing-sensitive-data-study-finds.htm?itc=xbodyrobwes [Accessed 14 Feb 2017].

Winter, A. (2015). *The Dark Net isn't what you think. It's actually key to our privacy*. Available at: https://www.youtube.com/watch?v=luvthTjC0OI [Accessed 14 Feb 2017].

Zlawdon, D. (2015). *Facebook and Google are cashing in on our secrets*. London, UK: Daily Mail Magazine.

14

Creating effective, timely, and valuable documentation reviews using a risk management framework

Annette Wierstra and
Joe Sellman

Abstract

Reviews are an integral and required part of the documentation process that serve to enable content experts to give their approval. However, the authors' experiences are that reviews are often thought of as separate entities to an overall project/process, with little or no rationale for what is needed. The authors contend that – as technical communicators – we can, and should, provide a useful and valuable assessment of documentation review processes at the planning, execution and sustainment phases of a project using a risk management framework to help focus and prioritize reviews. In this chapter, the authors will discuss their own experiences in documentation review, with the aim of providing guidance and best practices that other technical communicators can use to improve and influence reviews in their own workplaces.

Keywords

continuous improvement; document maintenance; reviewers; reviews; risk management

Introduction

Traditional texts in technical communication usually touch upon documentation reviews, but they often do so in a narrow scope, and are limited in the practical advice they provide. For example, the information often focuses on the value of reviews, the types of reviews and reviewers, and how to manage a specific type of review. While existing literature relating to reviews in technical communication provides a good starting point for anyone new to the topic, it is our belief that there is a sizable gap between the information provided in the literature and the practical knowledge required to successfully understand, plan and manage documentation reviews. Our main contention with the traditional view of documentation reviews is primarily that reviews do not exist in a vacuum and are only to be considered during the editing stage; reviews are instead part of the overall documentation project. As such, reviews should support the goals of the project, and reviews should be considered at all stages in the project: defined in the planning stage of documentation, executed as the documentation is developed and completed, and revisited in the continued maintenance of the document. We also argue that reviews should be reframed not only as a method of validating content but also as a tool for addressing project risk.

In this chapter, we aim to use our own experiences coupled with information from outside the traditional technical communication field to provide a risk management framework for effectively planning, managing and capitalizing on documentation reviews to provide value in the workplace. The chapter will look at a key, often overlooked, aspect of documentation reviews: the rationale for reviewing and how to plan a review process that adequately weighs risks and the needs of the project. This will be supported with practical advice for how to manage the documentation review process and how to use this rationale to set up a review process that will get valuable feedback, and what to do with that feedback. Finally, the chapter will consider the sustainment of documentation in the workplace. This chapter will show that carefully planning and correctly managing the documentation review process can enable a workplace to maintain good employee relations and encourage a dynamic and risk-averse environment. Conversely, it will show how getting the documentation review process wrong can negatively impact a business.

Literature review

It isn't hard to find a satisfactory definition of a review in technical communication literature. For example, "Reviewing means evaluating how well a document connects with its intended audience and meets its intended purpose" (Lannon, 2006, pp 109-110). Although reviewing documents between writers, management, and technical experts is a "pervasive, iterative, and common organizational process" (Kleimann, 1993) that reaches beyond the scope of technical communications, it is not unusual for the literature to offer only a cursory paragraph describing how to do a review. Often this paragraph only covers the practical side of editing your documentation, providing advice that recommends we methodically check our documentation content for accuracy, clarity, organization and effectiveness (Ewald, 2014; Lannon, 2006). We are advised to do this with a team, assigning each reviewer the task of editing for technical content, style, or suitability for the audience (Ewald, 2014). This team can include reviewers who are not only writers, but also subject matter experts, users, or designers who can provide a broader view of the documentation to better suit the project's goals and intended audience (Hargis et al. 2014). This is good advice that provides a solid foundation, but too often this is where the literature stops.

That is not to say that there is no literature that provides technical communicators with more specific advice about review methods. Barker (2002) outlines a practical six-step process for documentation reviews: review objectives, determine the reviews needed, establish a schedule, plan the reviews, send instructions, prepare material. Additionally, Barker provides insights into the importance of setting expectations, the purpose of reviews, and how reviews differ from testing and editing. Hackos (1994) indicates the importance of reviewers understanding the context within which they are reviewing. This includes the content specifications that the documentation is based on and where the document fits in the overall structure of information being provided (1994, pp 309–310). Providing expectations and context to reviewers allows them to provide feedback about the documentation with a purpose and insight that better aligns with the needs of the technical communicator.

Although there are improvements in information technology that can help technical communicators manage the review process, the challenge is that it remains primarily a manual process. Even if intranets and documentation review tools reduce the paper trail and improve traceability of our reviews (Ray, Ray and Downey, 2001), ultimately the process requires us to read, assess, and determine how to implement our feedback. Swarts (2004) did a study on mediating reviews comparing solitary review, reviewing texts in person, and using a textual editing replay during an in-person review. Although he found that using a screen capture replay did facilitate conversation and understanding between the writer and reviewer, Swarts also discussed the technological challenges of working with a tool that was not designed to facilitate the writing and review processes. Even looking at the available tools, technology has made it easier for us to share, change, and manage files with our reviewers, but it has not replaced the technical communicator's role in integrating that feedback into the documentation.

Advice about how to ask for feedback and implement edits into the document is helpful, but our greater challenge is in managing the complex relationship between technical communicator and our review partners within the context of organizational culture. Kleimann (1993) recommends a collaborative review process rather than a hierarchal one because this encourages thoughtful negotiation and incorporation of ideas and suggestions. The hierarchal approach creates a environment where reviewers are more likely to accept what is presented because critical review is not encouraged. A collaborative approach creates an environment where reviewers and writers can learn from each other and fill in each other's knowledge gaps with specialized expertise (Kleimann, 1993). Hackos (2007) also argues for collaboration between reviewers. The time spent planning, creating style guides, and tracking the team's progress is repaid in saved time because it reduces "duplication and overwriting" (p. 501) of content.

An additional challenge in document reviews is that the documentation must be reviewed not only for accuracy and completeness in its content, but also that it complies with the organization's actual practices and fits within the larger scope of the organization's documentation needs. This can be especially challenging if the writer is not familiar with the company culture and processes (Swarts, 2004). A deeper understanding of the organizational practices and culture can also help the writer and reviewer redirect these in a way that helps to implement positive change (Swarts, 2008). Reviews are also challenged by the gap between 'writer' and 'reviewer.' Usually reviewers are not professional writers but were sought out because they fill a gap in the writer's knowledge or provide expertise on the subject matter. This means that there can be a gap in communication or in expectations between the writer and reviewer (Swarts, 2004). This can be complicated because the nature of the review process requires that writers provide draft documents to their reviewers. These early drafts may be full of misunderstanding, incomplete information, or grammatical errors. The reviewers can be frustrated by these rough drafts if they don't understand the writing process and understand that later drafts will improve in quality. Framing the review process and setting realistic expectations for roles and stages of documentation development will help all parties understand the process (Hackos, 2007). Swarts argues that this can be further helped by "enculturation [which] implies that writers must learn how their texts function within the organization... [and] how they fit into the organization and contribute to its growth" (Swarts, 2008, p. 30). Collaboration and organizational context will facilitate deeper discussions and understanding between the reviewer and writer.

However, research has shown that these ideals for thoughtful, collaborative reviews that provide both macro and micro feedback on the documentation are hard to attain in practice. Usually the technical communicator is bound by "tight deadlines, multiple-reader review, simultaneous circulation for review, unspecified agendas for review, lack of consensus" (Swarts, 2008, p. 29), all of which reduce the quality of the reviews and the process. Feedback often defaults to composition, consistency, or whether the document looks and reads how a document should in the reviewer's mind. This is helpful, but doesn't get to a deeper examination of how this document fits into the organization's needs and work practices. The default is to review at the micro level, instead of the macro level (Swarts, 2004). This is amplified when the review is under a tight deadline; the review team will focus more on what is written than how the document needs to fit the larger project or organizational goals (Swarts, 2008).

In the early 1990s, Kleimann (1993) stated that reviewing is a subject that we know too little of and when we do look at it, research often focuses on the individual revision process rather than on how organizations can manage the overall review process. While reviews are a fundamental part of the technical communicator's work, it still seems to be one that we spend little time considering. The rest of this chapter will begin with a discussion about why we need to spend more time considering the review process in the context of risk management, then gives practical advice on implementing reviews in your workplace.

Reframing reviews within the organization's culture of risk

The writing process changes between writers and organizations, but the basics and commonalities are ingrained and ubiquitous. In our experiences, the writing process is usually a variation of the following stages (Barker, 2002; Ewald, 2014; Hargis, et. al, 2004):

1 Research: gather information, and determine scope of document.
2 Draft/refine: bring order, logic, and meaning to the information for audience.
3 Reviews: validate content.
4 Modify: adjust information based on feedback from review.

5 Sign off/approval: Approval that the information is correct/adequate for use.

6 Publish/deliver: Provide the audience with the information needed.

It is our contention and experience that writers see the first two stages as their primary concern (research and writing) and this is often at the expense of the later stages.

We believe that most people would agree that documentation review is essential. But why? How many organizations take the time to truly consider why they have documentation reviews? In our experience the short answer is usually: to validate that content is correct, applicable, and usable. However, rationales for reviews often assume three things which are not necessarily true:

1 Validation of content can be objective.

2 Reviews can be comprehensive.

3 Reviews are not bound by constraints of the environment in which they are conducted.

What do we mean by this? Well firstly, reviews cannot be objective, in that information is limited to the knowledge of the organization/reviewers/ industry at the time of the review. What is true today might not be true tomorrow, so documentation is either a snapshot of what was known and correct at a given point in time, or it is the start of a dynamic, growing body of knowledge that will require updating as more information becomes available and information changes. Secondly, no documentation and subsequent review can account for all possible situations; the information is limited to what is known and considered to be within scope. In the end, documentation is a tool, and cannot remove the need for the person following it to interpret and apply the information as needed. Finally, reviews cannot be an endless task. In business, limited amounts of time and resources are made available for documentation review and information validation. Think about how often a documentation project is held up due to the availability of key reviewers because these individuals' skills are in high demand in other parts of the organization.

In our experience, reviews often become derailed because of these assumptions. For example, we find that project sponsors/approvers (and many others) believe the three assumptions to be true, and in believing them do not address the reality of the review environment, leaving a lack of leadership and guidance on what the organization expects from a review. When these three assumptions are questioned, the review process must become more defined and dynamic. In short, it is our belief that the approach to reviews should shift the focus away from validating that the content is correct. Instead, *the review process should take into account the limited resources and time we have, and evaluate our priorities for review based on an assessment of the risk involved.* This shift in thinking is not necessarily an easy one to make, and it requires much more planning before reviews begin. However, it is our belief that changing to this pragmatic model reduces review cycle times, which in turn reduces the overall cost of documentation and can significantly reduce the time it takes to deliver a complete document.

So how does an organization plan the required reviews for a project based on this model?

A review plan requires a realistic examination of the available resources and the limits on these resources. All projects have limitations, whether these are available staff, money, or time. A large part of project management is risk management: helping organizations understand the risks involved in the various options available to enable them to make informed decisions about competing needs. Documentation reviews are similar to most projects in that they are commissioned, have goals, have resources allocated, have limitations, and are impacted by changes. It is therefore a puzzle to us why, for so long, the documentation review process has been treated differently, without the same focus on project management and risk management.

In his work *Risk Management*, Paul Hopkin (2013, p. 7) outlines the different components of risk management:

1 Risk agenda: why an organization conducts risk management activities.

2 Risk assessment: assessment of risks on finances, infrastructure, reputation and the market, and their potential consequences.

3 Risk response: an analysis of the existing risk controls and identifying the need for additional risk controls and contingency planning.

4 Risk communication: the creation of clear roles, responsibilities, and procedures relating to risks.

5 Risk governance: risk management with regular evaluation of existing and emerging risks.

Hopkin is looking at risk at the organization level, and while we are looking at a much more focused activity in documentation review, we can borrow significantly from his framework. The paragraphs below aim to demonstrate how we can elaborate on Hopkin's framework to apply it to a through analysis of risk in documentation review. When we examine the practicality of doing reviews through the lens of risk management, much of the process remains familiar, but it will change the extent of the review and focus on prioritizing according to this focus. This section will also revisit the common approaches to document review and fit them within this framework.

Risk agenda: fit reviews within risk culture

All organizations have a tolerance of risk, whether this is formally documented or whether it is just based on the workplace culture. Some organizations will have a low tolerance to risk, and others a high one. An organization's approach and focus of risk can depend upon the consequences of the risks. This relates back to Swarts' (2008) concept of the writer's enculturation: the writer must understand the purpose of the document within the organization's risk management approach. Technical communicators should review organization documentation and work with staff which will help them understand the culture around risk. Ideally, the stated values of risk and actual culture of risk within the organization are aligned. However, there are instances where they are not; for example, if senior management are trying to change the risk/safety culture within the organization.

Risk assessment: evaluate risk related to reviews

The risk assessment that should be considered for documentation should include:

- What are the potential consequences of information in the documentation being incorrect for the task?
- What are the potential consequences of no information being available for the task?

To answer these questions, it is essential to know both the nature of the work/tasks being documented and the people who will be conducting them (that is, a user analysis).

For example, if we took two hypothetical organizations at opposite ends of the spectrum of risk, a summary of the risk assessment might look like:

	Nuclear Power Station – Plant Operations	Image Software Company – User Instructions
Potential consequences of incorrect information	Nuclear meltdown	Users get stuck on a part of a task
Potential consequences of missing documentation (for example, an entire topic)	Nuclear meltdown	Users completely unable to perform a process

While this is a rather a simple example and does not consider the individual tasks and the risk associated with the different tasks, the point is to recognize that not providing any information can be just as risky as providing incorrect information. In the risk assessment, an organization should consider the benefits of providing some information rather than waiting for the information to be perfect. In the real world, a manufacturing facility might create a checklist for a task to fill a gap while the full documentation for the task is completed. This sort of assessment should help an organization to determine which documentation deliverables should be created first.

In practical terms, this means that when the documentation team understands the levels of risk of the documentation topic and the consequences to that risk, we have a benchmark to measure the review requirements. This allows you to focus your review on the tasks and topics with the greatest risk. Review requirements may be increased for new processes, products versus established processes, or products. The assessment should evaluate what kind of documentation exists, how well it functions, and what are the consequences of not having an in-depth review of that documentation. Updates to established, well-functioning documentation of a low- or medium-risk task or topic will not require as much scrutiny during review as a new documentation about new or high-risk tasks or topics. While this chapter is not focusing on planning the level of detail required for documentation, this same assessment can be used to decide the priority of what needs to be documented and the level of detail to which it should be documented. With this assessment complete, you will be better prepared to set review priorities and create a better plan for the review process.

Risk response: set review priorities

The risk response is the organization's set of controls for the various risks identified; therefore, it is essential that the risk response (such as the review process) is approved by the same individual who is the sponsor of the documentation project, usually some form of senior manager. The risk response to documentation is to conduct reviews, but this is not enough. As discussed earlier, there are limitations on what a review can achieve and the resources available for review. Using the risk agenda and risk assessment (identifying the risk tolerance and risk of tasks), we propose that the review process is customized to the risk. In our experience with reviews, there is a point of diminishing returns; generally the more reviews a document has been through the more likely the majority of issues have been found. However, documentation reviews tend to take the same amount of time regardless of how many have been done before.

Figure 1: Diminishing returns of reviews

Figure 1 does not have a specific number of reviews, or type of reviews (for example subject matter experts (SMEs), users, etc.) because the reviews considered by an organization will be based on their needs. The key idea of this graph is to counter the common belief that more reviews are always better. This graph is meant to represent our experience that at some point the additional reviews are not adding value, but when that point is reached for a particular document changes based on the risk assessment. This is subjective to the work environment because "risk management activities should always be proportionate for the level of risk faced by the organization" (Hopkin, 2013, p. 18).

Looking again at the hypothetical example of a nuclear power station, an operating procedure for a nuclear power station could include multiple SMEs and technicians reviewing the procedure multiple times during the process, whereas the instructions for a task in imaging software could be as simple as a single review from someone on the development team. These examples show the importance of understanding the risks associated in the workplace, and determining the number of required reviews based on this risk. The key is to plan a review process that balances the time and resources invested against an acceptable level of risk.

To create this balance of time, resources, and risk probably requires more than one review process. Within any industry, organization, or project there are varying levels of risk within documentation. Too often we have seen a one-size-fits-all review process for documentation, which can cause a large amount of unnecessary work or cursory reviews of high-risk documents. For example, in one of our previous workplaces all documentation went through the same review process meaning that the sign on the back of the bathroom door reminding staff to wash their hands should have gone through the same review process as a highly technical document. While this is a trivial example, it clearly highlights the need to right-size the review for the risk.

In an ideal process, all documents would be assigned some type of risk rating (or high-risk sections within a document would be identified). The risk rating could be used to assign documents to different review processes with more reviews and reviewers for the riskier documentation. One potential, but not necessarily unwelcome, outcome of such a system might be that there are a whole set of documents that are deemed to not require a formal review. For example, if there is only one person who does a specialized task and who is a qualified professional, you could let that person create their own documentation or use generic documentation that they customize to their situation.

This is a good place in the planning process to engage potential reviewers, which could be as simple as having a brief meeting. By engaging your reviewers early in the process, you can provide them with the larger scope of the project, allow them time to plan for the reviews, and engage them in the risk assessment of documentation. By engaging them in the full process and using their insight into which documents will require a more thorough review and approval, and which can be released after a cursory review or proofread, you take advantage of a diversity of knowledge and expertise early in the process (Kleimann, 1993). Time spent here will save time during the review stage (Hackos, 2007). In our experience, reviewers who are consulted early on are also more likely to provide a thoughtful and relevant review because they understand the project and our expectations. When we collaborate on deciding which documents really need in-depth review, the reviewers will also appreciate not having to waste time reviewing those low-risk documents. Additionally, planning the review stage may not eliminate the last-minute time crunch, but it will allow you and the reviewers to prioritize when the clock is running out.

Risk communication: define the review process

This is the stage in Hopkin's (2013) framework where the typical risk management documentation is found; that is, projects often have a risk register, risk plans, various procedures, and clear roles and responsibilities in the process. Tailoring this to our technical communication documentation framework forms two parts:

1 Communicating the review process and the strategy behind it.
2 Providing tools to those involved in the review.

We have observed in the review process that often individuals can be reluctant to sign their name to say that they have reviewed or approved a document. This can take the form of the approver asking for additional and unplanned reviewers and reviews. On reflection, we believe that this happens for the following reasons:

- The organization's assumptions about documentation review (as we outline above, that reviews should be objective, comprehensive, and not limited by the available resources).
- The organization's culture of when documentation is found to be wrong (such as after an incident).

For example, if an organization has a procedure for a task and, when performing the task, an incident occurs, organizations will often investigate to try to identify the root cause. The culture of the investigation is key. If the organization's culture is essentially to search for someone to blame, then our experience is that people who review/ approve documents will often delay a review process. If, however, the culture of investigation is to prevent the incident happening again, the reviewers/approvers tend to be happier to follow the documentation review process prescribed by the project. The type of documentation will help determine the risk, and the risk will determine when it is more appropriate to hold an individual accountable. For example, engineering is a profession where the engineer is supposed to sign off their name on a project for that very reason, to be held solely accountable. However, this level of individual accountability would not be required for all situations, for example an internal guidelines document for a software company.

If we have already engaged our reviewers and writers in the process of risk assessment and risk response, all parties will have a good understanding of where priorities and focus should lie within the review process. Reviews are improved when reviewers understand the context for their required feedback (Hopkin, 2013). This can be further emphasized with senior management backing these priorities and regular communication to the reviewers explaining how the review process has been developed to balance resources and risks, and the expectations and accountability of the reviewers/approvers. This approach to review enables reviewers/approvers to better focus their resources by indicating which documents (or sections of documents) are the riskiest and therefore where the reviewer should spend the most time. It also avoids them feeling it's necessary to undertake their own risk response, which could be determined by their desire not to be singled out if there is ever an error (which is likely the opposite of what is needed) or simply rushing through reviews more quickly as a deadline approaches. Communication also needs to flow back to the writer, so that together writer and reviewer can manage any time or resource crunches based on the already established risk response rather than on a random order of which documents are ready for review.

The second part of risk communication comprises the tools provided to those involved in the review. These tools usually build on the goals and priorities for reviews that we have already established. However, there will always be challenges. As research shows, the gap between the ideal and the practical is often large, especially when it comes to review (Swarts, 2008). Always partner the established goals and priorities with a practical process led by a designated person (usually the writer) who will oversee and manage reviews including timing, collating feedback, and confirming that the document is getting a proper review.

We always provide information with every document for the reviewer. This information will include a deadline, a contact, and an overview of the purpose or tips for doing a review. This reinforces the parameters set out during planning and helps the reviewer target their feedback towards the key issues. Our experience aligns with Swarts (2008), who finds that reviewers' default is to proofread rather than do more macro-level reviews, even if the review purpose was previously discussed. Remind your reviewer of the review goals by providing clear parameters. This will save both the writer and the reviewer time. One of our clients, when asked to review a short glossary of around 30 words for a specific document, came back with a self-created 1000-word glossary, months late, that didn't support the original intent. This creates an awkward situation of having to reduce hard work to a more directed tool for the document without devaluing the role and efforts of valuable team members. This could have been avoided with clearer communication about review expectations.

Setting deadlines on top of a list of parameters can feel like an imposition on a busy reviewer, but failing to set deadlines creates the false impression that reviews are unimportant and can be avoided or done quickly when they have time or at the last minute. It also gives the writer a timeline for follow up and to find ways to work with a reviewer who is struggling to complete the review. If the initial plan for solitary review fails to produce results, we find that flexibility can help to carve out time for the reviewer. In-person reviews can be more time consuming, but it can force all parties to engage with the document and provide more comprehensive feedback in a single review because the writer and reviewer(s) can ask and answer questions for clarity, which can make it more efficient than it may appear. We find that being aware of different communication styles of reviewers can also help; knowing if someone prefers to write or prefers to talk things through can help us pick the right kind of review for the different reviewers.

We have both used sequential and concurrent reviews (Kleimann, 1993; Swarts, 2008; Barker, 2002) or combinations of both to distribute reviews. Both methods have their strengths and weaknesses, so we prefer to use whichever works best with consideration to timelines (concurrent is faster), company culture, and the types and number of reviews. If there is time for consolidated comments (Kleimann, 1993) and further review, this can be a good way to merge feedback before doing a final review with more than one reviewer. Deciding which conventional review tools works best for a particular situation is part instinct, part understanding of the communication styles of the reviewers, and part trial and error, but being flexible rather than rigid will build better relationships with the reviewers and help keep avenues of communication flowing. The goal is to find a method that works both for the reviewers and the document in a way that limits endless document cycling and produces effective feedback.

Challenges can arise once the review feedback starts coming in, especially if there is more than one reviewer. Ideally, "document review is a joint action in which writers and reviewers build a common understanding of the text" (Swarts, 2004, p. 332). We find that common challenges are competing comments from multiple reviewers and differences of opinion between a review and writer, or between different reviewers. It is best to address these differences head on and not allow them to dissolve into battling comments, which only increases the number of review cycles and creates animosity. Ultimately, these situations can be solved by having clear parameters around who makes the decision about what to include. This is not necessarily the same person in all situations, as often this responsibility can be given according to roles; for example, writers have editorial control, subject matter experts control the technical content. It may also be helpful to return to the concept of risk: will the inclusion/exclusion of the information affect safety, accuracy, or frustration of the intended audience? Refocusing on risk and audience may help differing parties to reach consensus. It can also help to keep a portion of the document open for review and close the review of non-controversial sections to expedite the final reviews. Often face-to-face discussion and negotiation will help, but if it doesn't,this will require someone (perhaps the writer or a senior manager) to take responsibility for a decision.

Once the reviews are completed, the writer, possibly in conjunction with the reviewer, should evaluate the reviews against the initial plan. Are the reviews sufficient? Did we perform the right kind of reviews based on our review plan? Did other elements arise that may change determination of the risk rating for the document? Were all our questions answered? Were the questions/content addressed as we needed? It is good to be reflective about the review process, but keep in mind our overall goals. We are aiming for documentation that is good enough and meets the priorities we established in our plan, which may mean the document is not yet perfect. This means that our review process is not yet finished, but will continue with ongoing, planned document maintenance.

Risk governance: review documentation for continuous improvement

When a document has been approved by the required approver(s), it is complete. A job well done, we can all sit back and relax.

Except we can't. As we discussed earlier regarding planning documentation reviews, we believe that:

1 Validation of content cannot be objective and complete.

2 Reviews cannot be comprehensive.

3 Reviews are bound by the constraints of the environment in which they are conducted.

We believe that building in an expectation of document maintenance, or the 'living document' will relieve the pressure in the review stage because the expectation is not perfection but 'good enough' for the expected risks for the intended audience. Taking this approach, however, must be balanced with a systematic approach to continuous improvement in documentation through ongoing reviews. Risk governance within our technical communication risk framework addresses this in two ways:

1   The continued analysis of the review cycle.
2   The continued maintenance of documentation.

The continued analysis of the review cycle should use feedback from stakeholders to evaluate whether the review cycle is meeting the needs of the business. For example, if there are a number of incidents in a workplace relating to users following the instructions around heavy lifting, it should be considered whether the tasks are riskier than initially thought and perhaps the review process for these documents should become more thorough. The continued maintenance of documentation requires the same relationships and channels of information that we have already discussed for documentation reviews.

Once released, your documentation user becomes part of your review and testing team. You can expect users to find errors with the document, and these could be rated for risk as:

- Serious/urgent: an instruction that puts the user in harm's way.
- Moderate: there might be a couple of missing steps that the users can work out what to do, but the document keeps slowing down/ frustrating the users.
- Minor/non-urgent: a simple spelling error or an image that isn't entirely clear and could be improved.

While many users don't report any required or suggested changes, issues within documentation can make them feel frustration with the documentation or the process/product. When suggestions do come in, it is often without an indication of the importance of the change (for example, an incorrect spelling is as wrong as missing steps in a procedure). It is important for us to collect, collate and evaluate the feedback as it comes in. Low-risk changes (spelling errors) can be tabled until a later update, but if serious errors are found (missing steps), the risk involved may merit other methods of document updates and addendums.

Documentation must also be reviewed due to inevitable changes to the documented process. For example, processes are often modified to increase safety, reduce time taken or use less material, or new equipment or software is adopted, requiring a rewrite. Alternatively, many organizations have a policy of regular, predetermined and scheduled reviews of the document by the document owner(s)/approver(s). Any such scheduled review plan should also be set based on the risk assessment and response. That is, review times should be set at intervals that match the risk of the document. For example, high-risk documents should be reviewed more often than low-risk ones. An organization could consider planning reviews based on the age of a new document and on what each scheduled review finds for example, after 3 months, 9 months, 18 months, then every 24 months.

Most organizations will also need a maintenance plan that balances the development of new documentation and updates to existing documentation. Too frequently, maintenance reviews are a quick rubber-stamping exercise: the documentation is working well enough and there are other more pressing projects. Often, creating new documentation is valued more highly than reviewing existing documentation. This can unfortunately mean that existing documentaton is never, rarely or inadequately updated, resulting in many documentation projects losing their value before it was fully realized. Our entire goal, in this chapter, is to help organizations avoid the outcome of expensive documentation that went through too many reviews, took too long to complete, has alienated the users it is intended to support, and is not updated and therefore cannot be relied upon. It is our belief that documentation reviews based on a risk management framework should also ease the burden of ongoing document maintenance. This is achieved by:

- Planning: incorporating the need and risk for document updates/review in the business.
- Implementation: through engaging and encouraging ownership with the end users of documentation.

This sets the framework for continued success in encouraging users to report issues, and for reviewers to have a better-defined scope of their continued documentation maintenance activities. Ideally, this review process could follow a similar one to a review of the new document. Review the risk, choose the types of reviews that suit the level of risk, and then do a review implementing all issues that were logged for update later to the document owner for approval of what to include.

In an ideal world, it would also be helpful to have a central repository for users of the document to report issues, and where they can see the status of them. As this feedback is collected, use a method to triage feedback for the document, leaving spelling mistakes until the next annual review and addressing a dangerous error in an instruction much sooner. Provide feedback to those who raise issues, acknowledging their feedback and indicating what happens next, whether it will wait until the next scheduled review or will be dealt with sooner. These items are intended to improve communication between the users, reviewers, and technical communicators.

Projections for the future

It is our hope that technical communicators, senior management, and project managers will continue to work together to *right-size* review processes and practices by using risk management tools. This would involve conducting reviews efficiently and effectively by encouraging different approaches for different document types and not assuming that all documents require the same review process. This refined review process will enable the reviewers to feel ownership of the documentation, and ensure there is an effective method for them to be involved in updates/improvements. It would also ensure *complete* documents are regularly reviewed/maintained (again, based on need as per the type of document) to continuously improve documentation to promote safety, accuracy, and relevance for the intended audience.

Conclusion

Reviews, when they are done well, require a lot of time, not just from technical communicators, but also from the reviewers. Our reviewers are often people who have been asked or volunteer to take this task on in addition to their usual work load, so it is important to treat their contribution with respect and value the time that is put in. It is our belief that documentation review is just not about validating content; instead, it is fundamentally about managing risks. We also believe that in using risk as an assessment tool to evaluate and prioritize the document with high risks that requires in-depth reviews and reduce or simplify the reviews of low-risk documents, reviews will become more efficient and the reviewer's attention can be directed where it is most needed. This provides all project members with a tool and benchmark to establish where review time is best served. Considering reviews right from the project planning stages, using good communication to indicate the context, scope, and type of reviews, and establishing a document maintenance plan that reflects the risk of the document will prove to save time and money. It will also create a review culture where the process is valued because it isn't one-size-fits-all but is tailored to the requirements of the document. This will further promote good reviews and better documentation.

References

Barker, T. T. (2002). *Writing software documentation: A task-oriented approach* (2nd ed.). New York, NY: Pearsons Education.

Ewald, T. (2014). *Writing in the technical fields: A practical guide*. Toronto, ON: Oxford University Press.

Hackos, J. T. (1994). *Managing your documentation projects*. New York, NY: John Wiley & Sons.

Hackos, J.T. (2007). *Information development: Managing your documentation projects, portfolio, and people*. Indianapolis, IN: Wiley.

Hargis, G., Carey, C., Hernandez, A. K., Hughes, P., Longo, D., Rouiller, S., et al. (2004). *Developing quality technical information: A handbook for writers and editors* (2nd ed). Westford, MA: IBM Press.

Hopkin, P. (2013). *Risk management*. London: Kogan Page.

Kleimann, S. (1993). The reciprocal nature of workplace culture and review. In R. Spilka (Ed.), *Writing in the workplace: New research perspectives* (pp. 56–70). Carbondale: Southern Illinois University Press.

Lannon, J. M. (2006). *Technical communication* (10th ed). London: Longman.

Ray, D. S., Ray, E. J. and Downey, R. (2001) Using an Intranet to facilitate document review: An informal case study. *Technical Communication*, 48(4), 514–525.

Swarts, J. (2004). Technological mediation of document review: The use of textual replay in two organizations. *Journal of Business and Technical Communication*, 18(3), 328–360.

Swarts, J. (2008). *Writing review, enculturation, and technological mediation*. Amityville, NY: Baywood.

15

Training technical communication students in structured content using DITA

Nolwenn Kerzreho

Abstract

The University of Rennes 2 added DITA training to its programme in 2009. From writing topics with a text editor to collaborative migration projects, the programme has evolved significantly over time. Teaching student technical writers is, of course, very different from professional training. The younger generation have fewer bad habits to shed but need to comprehend the business drivers behind the publishing of technical content. Students are also at ease with new technologies. Nevertheless, there is considerable ground to cover to convince them of the industrial and professional usage of information within organizations and the business-to-business (B2B) world at large.

The students in University of Rennes 2 are from a multilinguistic background and not from a technical background. They usually have no difficulty creating new information in a structured writing setting while using the DITA standard. However, migrating legacy content to DITA means calling the shots on the tags and content, which requires a much deeper working knowledge of the standard. Working collaboratively on structured content with writers of different experience, which we mimic in a project mode in University of Rennes 2, proves even more challenging and rewarding for students.

This chapter draws upon eight years of teaching DITA that enables students to leverage their skills in the job market immediately after their training. It also is the result of discussions with technical communication practitioners, documentation managers, experts and teachers during international events, of surveying technical writer job postings in Europe and of trends in different industries over the last 10 years.

Keywords   higher education; minimalism; DITA; project-based training; professionalization

Introduction

What is most interesting from an educational point of view is the openness of DITA. As an open standard, it is available, free of charge, with no barriers, technical or otherwise, for research, learning, recompiling, extension and so on.

A lot of technical writing best practices from research and experience – from a wide variety of industries – are also baked into the standard, making it a perfect educational tool for teaching would-be technical writers. The standard covers most features that are included in proprietary tools and other standards, such as structured topic-based writing (Schengili-Roberts, 2016), metadata, reuse and conditional publishing.

The advantage for the students – compared to learning how to use specific software – is striking: the skills, the mind-set and the techniques are all easier to learn through practice and more transferable to other related techniques, tools and editors.

Today, we observe that trained alumni from University of Rennes 2, in France, work as information professionals in France, Germany, United Kingdom, Switzerland and Italy. It seems the shift from a linguistic to a technical skills set is becoming more and more tangible in the area of structured content.

The author would like to take the opportunity to thank the following people for their support to the programme:

- Kristen Eberlein from Eberlein Consulting and Chair of the DITA Technical Committee for her lectures to the students.
- Dr. Tony Self of the University of Melbourne and director of HyperWrite for his lectures, his help regarding the DITA programme and his useful book the *DITA Style Guide*.
- Ian Larner, from IBM UK, for providing the university with two information-packed workshops.

- All the people included in the OASIS TC Committee for their outstanding work.
- And last but not least, Marie-Louise Flacke, from Awel-a-Ben, the trend-setter who started the specialization in University of Rennes 2.

A European context at University of Rennes 2

The Master's degree programme in translation at University of Rennes 2, France, is officially titled: 'Master's degree in translation–localization and multilingual, multimedia communication'. It is run by the Training Centre for Technical Translators, Technical Writers, Terminologists and Project Managers.

This programme is a member of the European Master's in Translation network, and coordinator and project leader in the European project 'Optimising Translator Training through Collaborative Technical Translation', both aiming to establish a high-quality framework for advanced education in technical translation studies. These two projects are supported by the European Commission's Directorate-General for Translation.

The curriculum in technical communications can either concentrate on localization or multimedia communication specifics. Although technical writing has been part of the curriculum since 1998, several major changes imposed a stronger specialization in the last semester of the Master's degree. In 2008, the degree was renamed the 'Master's in Multilingual and Multimedia Technical Communication'.

Moreover, several European-wide projects, such as the Thematic Network Policy (TNP-3) in the area of languages, were determined to coordinate policies between academic institutions and other stakeholders designed to "overcome the frequently observed disconnection of higher education programmes from the needs of the non-academic environments and from research" (TNP3, 2013). Their findings pointed to the emergence of technical writing tasks in primary technical translator and project management work.

The Master's degree includes two mandatory work placements each summer and collaborative projects throughout the years, mixing students from the two years. All students must have three working languages, one of them being English.

Table 1: Evolution of the technical writing teaching in the programme

Timeline	Changes to the curriculum
1998–2004	Recruitment of a part-time technical writing teacher specializing in technical communication.
	Covers the third year of the Bachelor degree and the two years of the Master's years.
2004–2008	Technical communication is introduced as an option in the Master's degree.
2008–2012	Heavy specialization starts during the last semester of the second year of the programme.
2012–2016	The new syllabus includes a streamlined specialization covering the last two semesters of the Master's degree, followed by a six-month work placement.
2016–2020	The syllabus now expects the specialization to start in the first year of the Master's programme, thus spreading the in-depth learning of technical communications over two years.
	Ideally, the two work placements take place as technical communicators / information designers and conclude each year.

Why teach tag-based languages?

Tag-based languages, such as HTML and XML, have been taught in the translation programme for over 15 years. These tag languages are heavily used in the software and the video game localization industries, where content includes the properties of string files from user interfaces, subtitles from the software itself and so on. The tags are, of course, used in the web world, with HTML and scripting. Translators and communicators alike need to understand, extract text, translate and test tagged content. As it happens, students from University of Rennes 2 are recognized and hired in the gaming industry.

- Tag-based languages are already familiar to most students as they are heavily used within software development and in web localization work.
- Tag-based interfaces are also familiar to most students as they are heavily used within translation workbench software.

Tag-based interfaces and content are therefore relatively easy to teach as most of the students would have already encountered them through format training, collaborative projects and during their work placements.

Why OASIS DITA?

A report from the Thematic Network European Project (TNP project) in the Area of Languages (Université Rennes 2, 2008) mentions Darwin Information-Typing Architecture (DITA) as being part of an ongoing development of 'authoring' solutions, especially when combined with translation systems. DITA is also the most complete XML architecture readily available today. We have considered other options but dismissed them, as DITA has proven to be the superior standard in terms of both flexibility and features.

The Organization for the Advancement of Structured Information Standards (OASIS) is a non-profit, international consortium that creates interoperable industry specifications based on public standards such as XML and SGML. Although DITA is not the only architecture to use several modern content creation features (OASIS Technical Committee, 2015), it does cover most of them. These are modern ways to create content that can be found either as best practices in technical documentation methodology, such as structured writing, or as practical options to build documentation in proprietary tools, such as conditional publishing. The main features include:

- **Structure** – Most modern methodologies and toolchains embed structured content and XML. Although structured writing can be done in non-XML environments, with a desktop publishing tool for example, a constrained structure makes the standard easier to apply. DITA provides simple and industry-agnostic structures that the students must follow.

- **Topic-based writing** – Information typing probably originates from the Sequential Thematic Organization of Proposals (STOP) (Tracey, Rugh and Starkey, 1965) technique established at Hughes-Fullerton (aeronautical industry). It is also a recognized recommendation for Simplified Technical English and in DITA originates directly from the IBM guide (prior to web content) (Schengili-Roberts, 2015). DITA offers topic templates which are easily understood by the students.

- **Modularity** – Modularity is essential to understand topic-based writing and the flexibility it provides in terms of topic reuse in different documents. DITA includes maps, which organize the hierarchy of the content included in the topics. The flexibility offered by maps provides the students with the ability to test new information architectures at will.

- **Metadata** – Metadata are essential to the identification, categorization and control of documentation sets. They are required by most tools and leveraged by taxonomy practices when organising the information. Moreover, metadata are at a premium for the most modern web delivery tools and a very important skill for the students.

- **Reuse** – Reuse is a central concept for modernizing documentation. It enables more consistency and lowers the cost of producing updates and multiple documents with the same source content. Reuse also lowers the workload required from reviewers and translators. DITA offers extensive reuse capabilities through topic reuse, fragment reuse, indirect reuse with keys and variants.

- **Conditional publishing** – Conditional publishing is an alternative to creating two different sets of content, each applicable to different contexts of products. DITA includes tagging and filtering options along with the ability to create filters on variants.

- **Syntax** – Without going into too many details, the syntax in DITA follows a lot of the best practices students encounter in their technical writing classes. For example, hazard statement structures, the ability to add short descriptions and metadata, and the troubleshooting topic types.

Moreover, DITA offers these significant advantages:

- **Open standard** – The architecture is fully documented. The specifications themselves are publicly accessible online at no cost. Moreover, the first chapter of the specifications includes the basic concepts of DITA, topic-based writing and maps.

- **Choice of editing tools** – The architecture is an open standard, so there are multiple editors to choose from: text editors, XML editors, DITA-specific editors.

- **DITA Open Toolkit** – The DITA Open Toolkit is an open source project. It is made available to the students in the classroom and labs. The students can therefore test different options and run builds to produce their own content in the output formats.

As an open format, the students can easily access the specifications, check the document type definitions (DTDs), test and publish content, all without being tied to a specific tool. As the architecture enforces best practices, it is a good complement to other technical communication teaching, such as minimalism, hazard statements writing, instructional writing, user-centred writing and web writing.

Project-based training: taking learning a step forward

Learning context

The approach to learning DITA in University of Rennes 2 has several specificities. The technical communication specialization is deeply rooted in the Master's overall translation programme. University of Rennes 2 is a leading university for translation managers, technical writers and terminologists. The Master's programme is part of the *European Master's in Translation* network, a quality label for programmes. The students gain prominent project management and localization skills on top of terminology management and software skills.

The programme also emphasizes practical learning, or "learning by doing" (Carroll, 1998). It is no coincidence that the technical communication programme, where minimalism training is also provided, focuses on practice rather than on a theoretical approach.

The Bachelor-Master-Doctorate system lends itself to an incremental learning. The first year, the Master's students learn about the basics of topic-based writing and structure and how to apply DITA tags to content. They first use samples, then create new content, then make a small conversion. They also have the opportunity to participate in DITA conversion projects piloted by second-year Master's students.

During their second year, the students do further content migration and collaborative work. Under the supervision of a teacher, they work on content conversion and write all enabling documents to do so, including: content analysis, recommendations, style guides, templates, peer reviews and evaluations. They can pilot a project with volunteers from the first year. In the second year, the students also learn how to pilot a project, set up standards for review, provide guidance and templates and evaluate the work to their more novice counterparts.

The objectives are very practical. Each year, the students have to participate in a work placement consisting of four to six months for the first year and six months for the second year.

The curriculum's overall objective is to enable students to:

- **Year one**: handle DITA authoring projects as technical writers or ramp up to other structured content tools.
- **Year two**: on top of year one attributions, act as lone writers in a DITA environment or as junior information architects in organizations that already use DITA or want to switch to DITA.

Introducing DITA writing complexity

The overall learning is incremental and covers the basics of technical writing. Regarding DITA, the students usually follow an introductory course, then an advanced course, and finally a team-work project with an internal or external work provider. The learning curve is detailed in Figure 1.

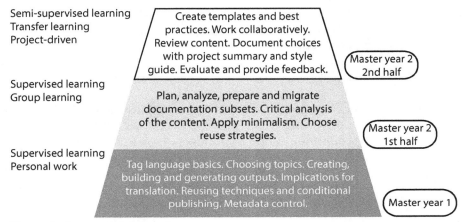

Figure 1: Incremental learning

The practical approach is described in Figure 2, from the introductory course (Year 1) to the heavy current specialization.

At the end of the introductory course, the students are able to grasp the basics of DITA writing:

- Organize a technical document subset into the appropriate number of topics and select the appropriate topic types.
- Adapt content for translation and topic-based writing.
- Create and reuse content.
- Publish to multiple outputs.
- Understand the challenges posed by structured XML content for translation (context, packages, XML strings).

After their first year, students undertake work placements mainly as junior technical communicators, whether or not DITA is used in the organization.

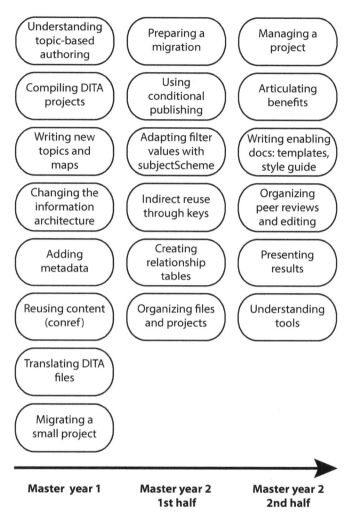

Figure 2: Practical orientation with example tasks of the training

Liaising with junior writers: writing in a team or batch writing

Due to its modular nature, DITA lends itself particularly well to collaborative work. Basically, as there are multiple objects easy to create and manage, the students can work simultaneously on different parts of the same project. The students can share the workload for information development and review, organize and reorganize the information architecture and share reusable content easily. Applying DITA in professional projects is especially appreciated by the students.

Reusing content that has been worked on by other people is quite familiar to translator trainees and students.

Content development workflow is straightforward compared to content translation – extraction of the text, as well as batch translation (Gouadec, 2010), are non-existent – and therefore easy to grasp for students.

However, collaborating with other writers, especially with junior students, proves to be very important as it brings a series of other questions, such as:

- **Writing conventions:** What tags to use? When? (style guide)
- **Naming conventions:** How to name topics? Maps? Images?
- **Organising topics:** How to share the same structure?
- **Templating:** How to best help the junior writers with templates? Examples?
- **Reviewing:** How to evaluate the junior writers?

The students also take on the role of reviewer, checking not only the content itself but its compliance to the templates. The students then provide feedback and adapt their ways to match the junior writers' skills. When there is sufficient time, a formal appreciation is sent back to the junior students.

The students also explore deeper and more complex mechanisms for reusing DITA-based content with keys, along with specializations and some CSS styling.

I believe that the project aspect is a very important part of the programme and is widely enjoyed by the students.

Delivery and liaising with 'customers'

When working on a project, either the company lead or a professor takes on the role of the work provider. The students are then asked to create a final document of their analysis and work delivery. The document also contains explanations and clarifications of the choices made.

This usually proves to be difficult for students as they have little knowledge of how to handle actual 'customers'. They sometimes lack confidence when presenting the results of their analysis to the practitioners who actually wrote the project or to project managers who manage information development in the organization. Plus, these new interlocutor's profiles are often seen as future potential employers by the students. However, this is a very rewarding experience for them and usually the project leads are quite happy with their discussion with the students.

Assessing the students' work

In correcting thousands of documents for DITA training, I created a scale to grade the student's work. The first level of expertise is principally based on technical points, whereas the second level of expertise assess the quality of the choices made in the content and focuses on the functional quality of the final deliverables:

- Technical evaluation: compilation, understanding of topics and maps
- Syntax: syntax rich, follows best practices
- Information architecture: structure and organization, linking and alternative navigation (related links), indexing
- Accessibility: metadata, image processing
- Organization and naming conventions: file handling and organization
- Specificities: conditional publishing, content reuse, indirect reuse
- Functional quality (migrating to DITA)
- Templates and guidance (project management).

The evaluation scale is sometimes shared with the senior students and enriched with their own project specifics so that they can evaluate the junior students' work.

Work placements and jobs

Some students have been working directly with DITA, others have been working with alternative content architectures, both custom or standard. None have reported difficulties. Other students reported using methodologies embedded in the DITA architecture, such as the topic-based authoring, even with non-XML editors. This confirmed knowledge transfer from one technology to another enforces the idea that DITA is a good candidate for learning structured writing.

Students who have worked with DITA and who have good project management profiles perform best with companies that are just starting out with DITA. Companies that welcome students during their work placements include large global companies with tools and processes in place, but also smaller, local businesses.

First jobs for graduates mainly include larger international companies based in France but also in other countries. The job titles range from technical writer to information developer, consultant and information architect. About one third of students work outside France (Kerzreho, 2012) mostly in Western Europe countries, including: Belgium, Switzerland, Germany, Spain, Ireland, Italy. In 2015, two francophone alumni were hired as technical communicators in the UK, thus making the point that technical knowledge can supersede language skills as hiring criteria.

Considerations on the syllabus

Reinforcing familiarity with standards

In parallel, the students learn about several technical communication standards. Standards-writing is a mandatory learning module that enforces familiarity with what is required in the DITA syntax choice. Actually, some proper writing constructions, such as writing hazard statements, are discovered first in an environment that is tool-agnostic.

Students recognize previous knowledge and how it is implemented in the DITA standard. Using DITA reinforces their knowledge of best practices. It also means that the students de-correlate the tool (XML editor), the architecture (DITA), and writing techniques. Foremost examples include: topic types described in Simplified Technical English, hazard statement writing using the ISO 3864 and ANSI Z535 standards, DITA web accessibility features with HTML writing, the need for short description with web writing techniques, the need for troubleshooting information from minimalism (van der Meij and Flacke, 2016). Of course, this reduces the cognitive load for students who can then focus on the DITA tools and other techniques, such as reuse.

Technicalities of the DITA Open Toolkit and XSL processing

There is a fine line between the DITA practitioners and the IT and tools specialists. In University of Rennes 2, the programme is still light regarding the technicalities of the DITA Open Toolkit (DITA OT) and XSL processing.

There were several reasons not to deepen this part:

- The information architect is not necessarily expected to become an XML/XSL developer.
- The person in charge of specializing the outputs for DITA is a separate role, as per the article describing roles and responsibilities for the DITA projects (Samuels and Bissantz, 2013).
- Teaching and training resources are currently still scarce, and the DITA OT evolves at a fast pace.
- The student profile does not necessarily match with required skills for the DITA OT development.
- Other profiles in the industry match these skills, including XML developers and engineers.

As of today, the students' feedback has never mentioned this. However, a few companies have required good stylesheet development skills.

Content management and content lifecycle

Structured content management is focused on how to effectively handle a large number of files mainly for collaborative writing, modification tracking, annotation, link tracking and leveraging metadata. The legal requirements from the industries are so diverse that it seems difficult to draw from a recognized set of knowledge. The theory related to the management of content life-cycles in both Waterfall and Agile methodologies and the link between content and product development were effective for the programme. So far, most practical learning of version control and knowledge management comes through the work placements. Interestingly, the students did not report this competence as missing nor companies as requiring this skill. This is probably linked to the multiplicity of DITA content management systems and the plethora of proprietary tools available today.

Assessing the framework

The DITA courses are, of course, only one part of the larger framework for the competencies in technical communication so it is difficult to separate DITA *per se* from the larger curriculum.

The endorsement of the curriculum's quality lies in the students' work placements and appreciations, the final approval being their employability.

Every year the curriculum is assessed and a student employability survey is conducted through several channels. The framework quality is thus assessed by three different partners:

- **The university itself** – through organizing and participating in academic and professional conferences, researches and contributions, curriculum adaptation, consulting publications from both academics and professional organizations.
- **The students** – through systematic training assessments, feedback after work placements on the appropriateness of the curriculum and feedback during their first three years after graduation.
- **The hiring organizations** – indirectly with work placement offers and skills demands, or more directly with students' assessments. The professors and adjunct teachers also look through competencies required and job definitions and liaise with the professional societies.

A survey conducted in November 2016 on a cohort of four classes of students from 2012 to 2016 (graduation was in October 2016) showed that 65% of the alumni found that DITA learning was an important to a decisive criterion in obtaining their first position: 63% were currently using DITA, 16% another open XML standard, 16% a proprietary XML, and 5% were using topic-based authoring techniques without an XML enforcement. The total of alumni using DITA and assimilated structure therefore amounted to approximately 82%.

Projections for the future

In over eight years of teaching DITA, implementing the standard, discussions with the students and guided project management, these are the main findings:

- Minimalism basics are *necessary* to understand what is involved in rewriting content for the DITA architecture.
- Learning a standard is not a replacement for critical thinking. Students must learn practical methodologies.
- Group work is excellent for analysis and confrontation of choices made regarding the information architecture.
- Guidance for junior student writers helps senior students better understand their own processes.
- Migrating content to DITA is the most effective exercise to challenge the students' practical knowledge of DITA.

The addition of DITA has tremendously enhanced the understanding of challenges and possible solutions regarding the production of documentation for the students. The practical and graduated approach to learning has proven to be the best so far.

The future syllabus includes more hours pertaining to technical communication, thus providing students with more possibilities to switch between the translation and the technical writing specializations. The programme will also include a Master's-wide project for collaborative technical writing, as a mirror project to the Optimising Translator Training through Collaborative Technical Translation (OTCT) translation project whose "aim is to enhance the integration of professionally-oriented practices in translator training curriculum" (OTCT, 2016). This means a week-long examination in simulated professional conditions. Of course, the academic team is also involved in European curriculum development.

Conclusion

A practical approach to standard structured writing has proved an excellent addition to the curriculum. Choosing DITA ended up being the right choice since the students can easily access the information, the professional expertise, and a broad choice of tools for practical projects is also accessible. Students can either work directly with DITA content, work with other standards more easily, move to other tools more readily, or even apply the underlying methods to desktop publishing, including structured writing, topic-based authoring and a minimalistic approach.

References

Bellamy, L., Carey, M. and Schlotfeldt, J. (2012). *DITA best practices: A Roadmap for Writing, Editing, and Architecting in DITA*. Boston, USA: IBM Press/Pearson Education.

Carroll, J. (1998). *Minimalism, beyond the Nurnberg funnel*. Cambridge, USA: MIT Press.

Evia, C., Sharp, M, Pérez-Quiñones (2015). Teaching Structured Authoring and DITA Through Rhetorical and Computational Thinking. *IEEE Transactions of Professional Communication*, 58(3), 328–343.

Gouadec, D. (2010). *Translation as a profession*. Amsterdam, The Netherlands: J. Benjamins Pub. Co.

Hackos, J. (2007). *Information Development: Managing Your Documentation Projects, Portfolio, and People*. USA: John Wiley & Sons.

van der Meij, H. and Flacke M.-L. (2016). What Research Has to Say About Supporting Error Handling in Training. *Best Practices*, 18(4), 94–98.

Kerzreho, N. (2012). Looking Back on Technical Communication as a Master Specialisation at University Rennes 2. *Proceedings of the European Academic Colloquium on Technical Communication*, Volume 1, 29–35.

OASIS Technical Committee Docs. (2016). *DITA Version 1.3 Specification.* Available at: http://docs.oasis-open.org/dita/dita/v1.3/dita-v1.3-part0-overview.html [Accessed 15 Dec. 2016].

OTCT project (2016). Available at: http://www.otct-project.eu/ [Accessed 15 Dec. 2016].

Samuels J. and Bissantz D. (2013) *DITA 1.2 Feature Article: Roles and Responsibilities of a DITA Adoption.* Available at: https://www.oasis-open.org/committees/download.php/50770/DITA_Roles_Responsibilities_final.pdf [Accessed 15 Dec. 2016].

Schengili-Roberts, K. (2016). *Don Day and Michael Priestley on the Beginnings of DITA: Part 2.* Available at: http://www.ditawriter.com/don-day-and-michael-priestly-on-the-beginnings-of-dita-part-2/ [Accessed 15 Dec. 2016].

Self, T. (2011). *The DITA style guide: Best Practices for Authors.* Research Triangle Park, N.C., USA: Scriptorium Publishing.

TNP3 (2013). *Thematic Network Project in the Area of Languages III.* Available at: http://web.fu-berlin.de/tnp3/ [Accessed 15 Dec. 2016].

Tracey, J. R., Rugh, D. E. and Starkey, W. S. (1965). Sequential Thematic Organization of Publications (STOP): How to Achieve Coherence in Proposals and Reports. *Journal of Computer Documentation*, 23(3):4,68 (1999). Reprint of a January 1965 article.

Université Rennes 2 (2008). *MULTICOM – Status quo report – France.* Available at: http://www.sites.univ-rennes2.fr/centre-langues/images/MultiDocs/Outcomes/WP2%20Needs%20analysis/Status%20quo%20reports/MULTICOM_Phase%201%20report_FR.doc. [Accessed 15 Dec. 2016].

16

Are AI writers capable of work in the current workplace

Jason Lawrence and Chelsea Green

Abstract   While human professional writers can protest that artificial intelligence (AI) cannot possibly write better than them, that doesn't change the fact that AI can write and edit. This feasibility project accepts an AI writer can be a coworker on a team of human technical writers. Therefore, based on current technology, is it feasible to consider AI professional writers as a future trend? The answer to that question requires a review of AI principles, a survey of current technologies, and a showcase of current AI writers – some of those requirements can be completed by current AI.

Keywords   artificial intelligence; AI writers; Feasibility Report; workplace; technology

Introduction

The simple truth is that artificial intelligence can write. More important than whether or not an AI technical writer can replace human beings is whether it is feasible that an AI writer can be a co-worker on a team of human technical writers. This Feasibility Report checks against three different levels of technical writer experience. The report uses the Society of Technical Communication's knowledgebase (TCBOK) to define terms and match standard criteria in the field. As of Spring 2016, AI can write in the workplace; it can perform many duties of the Technical Writer I level (see 'Different levels of technical writer' on page 314). This improves the efficiency of entry-level writers, supports professionals who write, and gives experienced writers more time to dedicate to specialized documents. Most significantly, the technology exists and professional writers must begin formulating best practices before project managers form best practices for them, based on profit margins alone.

Workplace situation

This Feasibility Report rejects the notion that some work functions, no matter how intuitive or empathic, are exclusively human. Unfortunately, nothing is beyond the capacity of appropriately programmed AI. This Feasibility Report is about job function, not sentimental connection to the human touch. This Feasibility Report acknowledges the unfortunate business precedent that the human touch is too expensive and quality isn't always worth the price. The same business decisions that terminate technical writing positions, regardless of the need for a writer, are the same need-based decisions that will invest in AI, rather than humans.

Of course, the size of a business determines the need for technical writers. This is because of both budget and the scale of operations. A large organization might have a team of technical writers who write materials shared by multiple departments. Some such businesses may employ several teams of technical writers. A mid-sized corporation might have only one writer because a single department may not need to share as much material outside the department. In fact, specialists who do not need to worry about visibility throughout the organization might still do some of the writing. However, smaller businesses must make a decision whether their specialists need a writer or whether the budget can justify a technical writer, regardless of need.

Different levels of technical writer

The Technical Communication Body of Knowledge (or TCBOK) is a wiki sponsored by the Society of Technical Communicators (STC). The following three bullets present three technical writer career paths detailed on the TCBOK wiki (Tcbok.org).

- For a Technical Writer I, 0–2 years of experience are required. They are supervised by others.
- For a Technical Writer II, 2–4 years of experience are required. They are expected to be multifaceted in the writing they can do.

■ For a Technical Writer III, 4–6 years of experience are required. They may end up leading others, suggesting that judgment is valuable.

In fact, TCBOK identifies four career paths. Of the four technical writing career tracks identified by the TCBOK, this report uses only the first three. The fourth career track involves managing other technical writers. The management of function-driven resources is not beyond the capacity of an AI; however, the management of human workers is not within the scope of this Feasibility Report. While logistical management and quality control are within the capacity of AI, at the time of this reporting, AI doesn't seem to have the capacity to manage the functions of human technical writers.

Technical Writers I, II, and III are the roles of human technical writers in the current workplace. Rather than conflate all AI technical writer job demands, this report seeks to explore the feasibility of each level, defined by the TCBOK wiki. In addition to making a more specific feasibility analysis, the career paths were useful in building the AI profiles that are the core of this Feasibility Report. Consequently, the feasibility of an AI writer varies with the different level.

Future trends

Joe Gollner (2011) writes about business analysis in technical communication practices: "In particular, I have been thinking about how the skill set of the effective technical communicator may be an important asset for organizations to tap as they are forced, in an increasingly systemized world, to formalize their products and processes." The most systematized businesses are currently implementing automated metrics, machine learning (ML) tools, and business intelligence programming. Those same businesses look for automation solutions to cut costs and increase efficiency. Professional writers will need to clearly articulate the value they add to a business in order to remain relevant.

While 'remixing' the results of a five-year-old survey for 2009, Barbara Giammona recommended that technical writers up their game in the workplace: "Technical communication needs people who are able to provide specific skills in areas like user documentation, tutorials, administrator guides, training videos, command line information, and more. No one person will be able to do all of these, and to do them well." Just as Giammona would recommend defining these specific skills as important aspects of technical writers' specialization, the emergence of the AI professional writer means there are resources for the less-specialized technical writing responsibilities.

AI can write well enough

This Feasibility Report is about whether AI professional writers will be in the workforce anytime soon. However, can AI even write? The answer is yes. On March 7, 2015, the New York Times released an online survey "Did a Human or a Computer Write This?" with eight writing samples. A computer wrote six of the writing samples. The Feasibility Report team reported that the biggest surprise was Survey Sample 5. The six writing samples are included below:

Survey sample 1:

> A shallow magnitude 4.7 earthquake was reported Monday morning five miles from Westwood, California, according to the U.S. Geological Survey. The temblor occurred at 6:25 a.m. Pacific time at a depth of 5.0 miles.

Two years ago, Ken Schwencke, a journalist and programmer for the *Los Angeles Times* created an algorithm by the name of Quakebot. Quakebot pulls data directly from the United States Geological Survey (USGS) template. When the story is created, it is both reviewed and published by a human editor.

Survey sample 3:

> When I in dreams behold thy fairest shade
> Whose shade in dreams doth wake the sleeping morn
> The daytime shadow of my love betray'd
> Lends hideous night to dreaming's faded form.

Swiftkey is a popular machine learning (ML) Android application by MIT PhD student J. Nathan Matias. ML still requires a human element to produce the text. The app has a Shakespearean vocabulary and uses an algorithm to predict what word the writer intends to type next. Due to ML, the app can write for a user, based upon the user's own vocabulary and messages.

Survey sample 4:

> Benner had a good game at the plate for Hamilton A's-Forcini. Benner went 2–3, drove in one and scored one run. Benner singled in the third inning and doubled in the fifth inning.

Quill is an AI sportswriter robot, created by Narrative Science. It can turn sports statistics into a natural language newspaper story. A human manually captures game events (a pop fly in baseball, a three-pointer in basketball, and so on) – much like taking notes. Quill will generate a written retelling of the game's events. Narrative Science claims that Quill's news report is as accurate as the data that human users provide it with.

Survey sample 5:

Kitty couldn't fall asleep for a long time. Her nerves were strained as two tight strings, and even a glass of hot wine, that Vronsky made her drink, did not help her. Lying in bed she kept going over and over that monstrous scene at the meadow.

This is from *True Love*, one of the first computer-generated works of fiction, printed in 2008. The *St. Petersburg Times* reported, at the time, that *True Love*, published by the Russia's SPB publishing company, was the work of a computer program, PC Writer 1.0 and a team of IT specialists. The 320-page novel is a variation of Leo Tolstoy's *Anna Karenina*; yet, the novel is worded in the style of Japanese author Haruki Murakami. It hit Russian bookstores in the same year.

Survey sample 6:

Tuesday was a great day for W. Roberts, as the junior pitcher threw a perfect game to carry Virginia to a 2–0 victory over George Washington at Davenport Field.

This is another sports news report from Natural Science's Quill. There are interesting patterns that show formulated writing but professional writers are not writing classical literature either.

Survey sample 8:

In truth, I'd love to build some verse for you
To churn such verse a billion times a day
So type a new concept for me to chew
I keep all waiting long, I hope you stay.

Phil Parker's formulaic auto-writing is what he calls econometrics. All the algorithms Parker did mimicked what economists had already been doing for decades. Phil Parker, an economist, doesn't really write his books but uses sophisticated algorithms that can pen a whole book from start to finish in as little as a few minutes. The secret is sophisticated programming mimicking the thought process behind formulaic writing.

This Feasibility Report

Artificial intelligence has been word-processing for decades but it has only started writing in recent years. In fact, AI is getting very good. With AI technical writers comes a difficult time when human technical writers need to adapt their role in the workplace. Rather than vehemently deny that AI could ever do 'my' job, this Feasibility Report accepts that AI executes remarkable writing and explores the feasibility of AI technical writers.

1 **Is it feasible for an AI technical writer to act as automated support for technical writers and professionals who write?** Level I technical writers typically have 0-2 years of experience in the field, and are interns or beginner technical writers. Level I AI technical writers act as automated support for technical writers and other professionals who write, providing more support than a word processor, and can create or revise almost any document. Current AI is programmed with the basic grammatical and contextual knowledge to complete almost any level I assignment. As it is supervised by others, the AI learns from the corrections that its supervisor makes. Level I AI technical writers are experts in their field, having access to an immense online database of information.

2 **Is it feasible for an AI technical writer to take a position on a team of technical writers?** Level II writers typically have about 2-4 years of experience. Level II writers are also multifaceted. Level II AI writers participate on a team of technical writers. The programming that they have is not very different from the level I technical writer. They are able to run multiple programs at once and access hundreds of online references in mere seconds. AI is able to check its co-workers work for inaccuracies and effectively generate original writing to support a team of writers. They are also able to create graphs, visuals, and documents specific to their area of expertise.

3 **Is it feasible that an AI technical writer can meet the writing needs of specialized professionals?** Level III technical writers typically have about 4-6 years of experience. Level III AI are able to meet the writing needs of specialized professionals. Unlike the other AI writers, the level III writer is programmed with the knowledge and experience necessary to set standards and take initiative. The AI is also able to

correct its team's documents, communicate with no error, and communicate in foreign languages.

Current state of artificial intelligence technology

There is a lot of philosophy and science fiction that frames the public conception of AI. However, this report seeks to show how feasible it is to have AI technical writers in the workforce. This section briefly presents the history of AI, the current technological capacity, an array of cutting-edge examples, and examples of current AI writers entering the market. The objective is to create distance from both philosophy and science fiction; the objective is to show how advanced AI really is.

History of artificial intelligence

The Feasibility Team used Kevin Warwick's *Artificial Intelligence* (2012) book to divide the history of AI into four historic epochs. Warwick wrote his book at the close of the 1990s; his concept of current AI was a collection of projections into the future. Therefore, if the epochs are cleanly split at the decades then 1990 is the beginning of the modern era and the publication of Warwick's book marks the end. For good or ill, the Feasibility Team did not want to set the criteria for a fifth epoch. Therefore, the modern era is from the 1990s until this Feasibility Report.

Early history of artificial intelligence

When Alan Turing asked the question, "Can a computer think?" the Turing Test was developed, and so sparked the birth and history of AI. According to credited writer Kevin Warwick, there is a strong relationship between computers and AI. It began in 1950, when Claude Shannon proposed the first program for a chess-playing robot. Then, in 1956, John McCarthy coined the term 'artificial intelligence' at the Dartmouth Conference, devoted to the subject.

The middle ages

Many researchers, such as mathematician Lofti A. Zadeh, set out to disprove claims based on perfect machines and limited 1960s technology. A lot of research was on the development of natural human language input in machines, rather than machine code. An example of this intelligence is Joseph Weisnbaim's ELIZA, who made natural human language conversation between man and computers possible.

The dark ages

Three main problems were found in this epoch that were not easily overcome. AI lacked computing power, AI did not understand natural language, and machines cannot think the same way people do. Without computing power to resolve research challenges, Warwick claims the philosophers were able to turn attention from technological development to philosophical argument – like John Searle's Chinese Room.

The artificial intelligence renaissance

Finally, progress was made to improve the strength and ability of AI, proving to skeptics that robots can think independently. McCarty led research developing 'Expert Systems' that can perform autonomously with information uploaded by human programmers. Those programmers would have to anticipate every conceivable variable the AI would need to perform its purpose. Therefore, an Expert System must be preprogrammed and constantly fed both the most current information and most up-to-date instructions. In addition, Edward Feigenbaum developed one of the first computer models of how people learn and subsequently an AI Expert System that performed abstract problem solving. Feigenbaum's abstract problems were excellent applications for his Expert System to solve because he already had the formulaic computer model for how humans solve problems. There were many other research successes that put AI back on track. Unfortunately, after two decades, AI's first impression had already been made.

1990s to the present

AI is currently used as a service tool, capable of learning, teaching, assisting patient-related situations, cleaning houses (Roomba vacuum), helping the elderly, used in electronic appliance (mobile phones, laptops, home phones, mp3, iPod, and so on). The year 2011 brought the voice-activated SIRI. AI is used in industrial applications, financial systems, and the military.

There are new obstacles in AI research. For instance, Expert Systems alone were not a practical way to build autonomous machines; such machines constantly required human input – not very autonomous.

Consequently, many programmers look to machine learning (ML) to fill the gap, while other researchers develop more adaptive AI. ML is a new method of programming that allows an AI service tool to store user preferences and performance results. Consequently, the AI learns to perform the job better.

The following are other examples of AI service tools that help represent the current applications of AI. Notably, these examples all show high technological capacity that can support autonomous AI, defying decades of criticism and doubt. In addition, these examples showcase advanced ML.

- The Neato XV-11 vacuum cleaner travels in straight, partially overlapped line, has sensors that prevent it falling off stairs, and uses an algorithm to map the room being vacuumed, before it returns to its home base and recharges itself.
- Robonaut is a humanoid robot that can use space tools and work in similar environments to astronauts.
- Arachno-Bot is a 3D printable spider robot built as an exploratory tool for environments that are too hazardous for humans. With the natural spider as the model, Arachno navigates spaces as a real spider.
- Baxter is an industrial, two-armed robot with an animated face. It is used for simple industrial jobs such as loading, unloading, sorting, and handling of materials, designed to perform the mundane tasks on a production line.
- Atlas is intended to aid emergency services in search and rescue operations. Atlas can drive a utility vehicle across rubble, remove debris, open a door and enter a building, climb an industrial ladder, use a tool to break through a concrete panel, locate and close a valve near a leaking pipe, and connect a fire hose to a standpipe and turn on a valve.

While these examples demonstrate incredible technological achievement, each AI is a service tool with only one preprogrammed purpose. On the other hand, modern AI is meant to move beyond preprogrammed purposes. After all, the autonomy of a machine is quite limited if a human has to update the coded instructions all the time. The objective is to achieve AI that can adjust and adapt when the purpose shifts, without necessitating human feedback.

Modern artificial intelligence

The AI service tools of the 1990s and 2000s don't need to possess free-thinking human intelligence to make need-based adjustments to performance. Rather, modern AI simply needs intelligence. Kevin Warwick is a researcher with a long history in the fields of AI and robotics. He states the key concept to understand modern AI; a machine doesn't need to have human intelligence in order to be artificially intelligent (Warwick, 2012). In fact, there is already a precedent; humans attribute intelligence to dogs, dolphins, monkeys, and the octopus. The history of AI is full of researchers and philosophers who set the bar at human intelligence. However, modern AI focuses more on the models of function-driven intelligence found in nature. For instance, an intelligence doesn't need to map and process 3D spaces in order to navigate an obstacle course; rather, an intelligence merely needs to navigate like an insect in order to traverse the same 3D space.

Basic neuron model

One way to imagine modern AI is to consider the neural system of the human body. This isn't to say that researchers have come back around to patterning AI after human intelligence; rather, the Basic Neuron Model is based on the relationships and behaviour of neural transmissions, neural receptors, synaptic processing, and the neural network of the system itself. This matrix of various relationships and behaviours is what researchers call *perceptrons learning*. The computer code manages programs, cycles, inputs, and algorithms according to some manner of neural perception. The end result is a Self Organising Neural Network. Rather than an Expert System, AI based on the Basic Neuron Model organizes itself, without external organization uploaded by a human programmer (Warwick, 2012).

Evolutionary computing

In an evolutionary model of AI, the AI grows in generations. Each generation is the result of what researchers call Genetic Algorithms. Much like the evolution of life, each generation optimizes specific biological mechanisms to respond to natural thresholds – subsequent generations may or may not evolve to biologically capitalize on those optimizations. The AI works in the same way; threshold levels, threshold values, threshold actions, inputs, feedback, and interfacing programs all contribute to an algorithm – or evolution. Once again, an Expert System doesn't process thresholds or compare itself against previous generations; rather, those thresholds are set parameters preprogrammed and unchangeable, without human intervention. However, an evolutionary AI will grow past those thresholds (Warwick, 2012).

Agent methods

Rather than a single AI, the Agent Method is concept of intelligence more along the lines of swarm intelligence. Multiple agents manage distinct aspects of a function – or many separate functions. The interactions of the agents can behave like either neurons or evolutionary generations; whatever the case, there is not a single Expert System to fix parameters on a swarm. Consequently, the agents engage each other as needed, outside whatever bounds may have originally been set by the human programmers (Warwick, 2012).

A final point Warwick (2012) makes is the importance of senses. He suggests that without sensory input "how can an entity perceive the world, respond, learn, or communicate" (p. 146). If researchers want to build adaptable AI then it needs to interface with inputs at a human level – perhaps even interface with humans. Such interactions are not beyond the capacity of the AI available now. For instance, iCub is a sensory robot (Figure 1). iCub has the "basic skills of [both] perception and data processing that might come hardwired in a human infant: visual, auditory and tactile senses" (Chet, 2009).

Figure 1: iCub touches an object

Though an older designed robot, the 2009 iCub is about the size of a three-year-old toddler that adapts to its environment, with a clear reliance on sensory inputs.

To illustrate the means by which AI can obtain sensory inputs, Warwick presented a range of technologies and organized them all according to corresponding senses (see Table 1). Our intent in presenting this information is the same as Warwick; we want to show the capacity of the current AI to both engage an environment and human beings. Warwick leaves out the sense of smell because without the need for human intelligence, taste and smell are redundant senses. Warwick also substitutes hearing with movement; insofar as movement and balance are regulated by mechanisms related to hearing, an AI need only focus on processing those inputs for movement.

Table 1: Four AI senses and the technologies that facilitate them

Vision	Movement	Touch	Taste
Image transformation	Triangulation	Material for touch	Ultra-violet detection
Image pixels	Active triangulation	Force sensing	X-ray
Image analysis	Sonar	Optical sensor	
Preprocessing	Radar	Infrared Ddetectors	
Image Sspectrum	Magnetic sensors	Audio detection	
Finding edges	Micro switch	Telemarketing AI	
Finding lines	Proximity sensors		
Template matching	Radio Frequency Identification Device		
Point tracking			
Threshold variation			
Segment analysis			
Forming segments			
Colour			
Image understanding			
Blocks world			

Five cutting edge AI

The following are additional examples of cutting-edge modern AI that use sensory input to engage both their environment and human beings.

Nadine

Figure 2: "The Social Robot" Nadine

Pictured in Figure 2, Nadine was created in 2016 by Naudia Talman and her team of scientists, at the Institute of Media Innovation. Avianne Tan (2015) of ABC news reports: "She smiles when greeting you, looks at you in the eye when talking, and can also shake hands with you". The robot is able to hold conversations, recognize people she has met, and is full of personality and facial expression. Her vocabulary expands as she interacts with people and she chooses what to say, making her dialogue with every person different. Nadine is able to argue back with a human along with an angry expression. In addition, she can draw and read stories to people who do not have a companion. The positive aspect about this AI is that she has the ability to learn as she continues on through time.

Sophia

Figure 3: "The ultra-realistic Humanoid" Sophia

Created by Hanson Robotics at SXSW Interactive in Austin, Texas, Sophia (Figure 3) is "based off of Audrey Hepburn" (Hanson). She has two sophisticated cameras in both eyes that are used to interact with humans. Also, she tracks facial expressions and eye movement of people and can recognize them too. Hanson uses a combination of Alphabet's Google Chrome voice-recognition technology and other software that helps Sophia process speech, hold a conversation, remember interactions and get smarter over time. In addition, Sophia has 62 different facial expressions she can use for non-verbal communication.

Amy Ingram

X.ai is a company that is centred on user scheduling. For instance, a user of this program would email either Amy or Andrew (depending if you want a male or female) their preferences such as "no meetings before 9AM" and Amy stores it in her memory, utilizing Google calendar and Gmail. Recipients can see what times or areas they can meet with the person. Company representative Stephanie Boyd (2016) remarks on Amy's efficiency, "I'm being commissioned to write a new piece of classical music by 50 violinists in 45 countries, and I'm meeting with everyone over Skype and Amy has organized every single meeting."

Enlitic

Enlitic is a medical company that has created a *deep learning* (a 'deeper' version of ML) technology that is meant to give patients the most precise diagnostic available. Unlike other medical diagnostic programs, Enlitic takes very detailed pictures of the targeted area and contextualizes the images and compares them to past images that have been taken, clinical data, and laboratory studies to make a diagnosis. Other diagnostic programs rely on assumptions about a particular disease and gives a broad decision and can lead to incorrect diagnostics. http://www.enlitic.com/

Self-driving car

One of the newest working AI that is in the media is the idea of a self-driving car. Companies such as Mercedes are in the process of creating a self-driving car concept. Before the idea of a self-driving car, some cars already had AI technology that parked them. More recently Volkswagen has implemented safely stop technology that actually applies the brakes for you if you get too close to another vehicle. To go even further Google has made a working prototype self-driving car that has been tested with the help of volunteers.

Enter the technical writers: cutting-edge artificial intelligence

Nadine, Sophia, and X.ai alone can illustrate what AI writers might look like in the workplace. Nadine can read the tone of a human's voice. By doing this, she also learns how to react correctly to the conversation she is in. Tone of voice is a major aspect of assessing subject matter expert needs. Since Nadine was programmed to begin as a blank slate, she picked up communication information from the team that created her, and eventually was able to communicate effectively with her team. On the other hand, Sophia was programmed to assess human facial expression to tell the mood of a conversation. In addition, Sophia's speech software includes Internet search capability so she is able to process information at a greater speed. Finally, X.ai has the ability to automatically email clients and schedule appointments; this contributes to the smooth running of the workplace. This automated emailing system can also supply customer service outside of work hours.

The real question is whether it is feasible that this modern AI, with the capacity to interpret sensory inputs, can be a technical writer in the workplace? The Feasibility Team found AI that can write. Beyond the writers described in the *New York Times* survey, mentioned at the beginning, the Feasibility Team found some great examples of writers.

- Emma/Mansi (machine augmented neural search interface): an AI built on top of a convolution neural network. In other words, she works for ETF Investing Industry, founded by Ben Fulton. Emma delivers financial business analysis, and she covers all stocks in the S&P500 (Standard and Poor 500). She wrote her own article about what ETF is and how it works. She is capable of writing an article in 20 minutes. http://etfdb.com/authors/emma-ai/ http://etfdb.com/news/2016/03/01/artificial-intelligence-writing-about-the-technology-etf-qqq/

- IBM's Watson (World Champion Jeopardy Computer) can analyze writing. Named "Watson Tone Analyzer", this computer program doesn't only do regular grammar and spell checks, it reads and interprets your writing and gives feedback on how the tone of the essay should be written and edited. http://www.computerworld.com/article/2949817/emerging-technology/new-ai-tech-helps-you-write-right.html

- Wordsmith is a program that takes published data (stats, sports scores, financial reports) and creates stories and written documents out of them. For every row of data entered, Wordsmith writes a story about it that is polished enough to be published and used as a personalized article. http://automatedinsights.com

- Quill, the "Advanced Natural Language Generator" is an AI program that doesn't just report the numbers for the data you are collecting. Quill can also interpret and create perfectly written narratives for all of the data. It'll even organize it for you and make it easier for everyone to read. Quill allows you to increase the value of your data by fulfilling the tailored information requirements of all audiences. Whether you are communicating to regulatory bodies, employees, business partners or consumers, Quill delivers 1:1 personalized communication in a consistent, brand-aware voice at a scale only possible with technology. https://www.narrativescience.com/quilll

- MIT was researching how they can use ordinary language so that a computer could interpret it and turn it into computer code. In a pair of recent papers, researchers at MIT's Computer Science and AI Laboratory have demonstrated that, for a few specific tasks, it's possible to write computer programs using ordinary language rather than special-purpose programming languages. http://news.mit.edu/2013/writing-programs-using-ordinary-language-0711

- Valentin Kassarnig of UMass Amherst created an AI machine that can write its own political speeches using an algorithm that can tell the probability of a topic being discussed within the category of the speech. He took about 4,000 political speech segments from about 53 US floor debates, with 50,000 sentences each of about 23 words. By categorising them as Democratic or Republican and using probability techniques, this AI has enough information to write political speeches that are similar to actual speeches (MIT Technology Review, 2016).

Feasability report methods

Feasibility project team

In March and April of 2016, 20 Southern Connecticut State University students took part in a feasibility project, as a significant portion of their Business Writing class. As a result of their feasibility research, they uncovered a lot of information about current artificial intelligence research in the field. Their combined efforts culminated in much more information than could be fit in to this Feasibility Report. For their research, 10 of the students wished to be credited by their name and their major of study (Table 2). We have called these 20 students the Feasibility Team. The following chart identifies the Feasibility Team, along with their majors.

Table 2: The names and majors of the feasibility team

Name	Major	Name	Major
Chelsea Green	Professional Writing, English, Co-Author	Alexander Bloomingdale	Accounting
Melanie Espina	Journalism	Dan Kingsley	Management
Shauna Plancon	Communication	Matthew Medeiros	Finance
Kaylin Tomaselli	English	Ted DeConne	Exercise Science
Shanice Tricoche	Nursing	Jessica Yupangui	Nursing

Another 10 students did not wish to be credited (Table 3). They came from the following specializations:

Table 3: Anonymous members of the Feasibility Team

Accounting 3x	Biology 2x
Education	Finance
History	Management
Nursing	

Feasibility team bias

The 20 students represent a broad spectrum of Southern Connecticut State University academic degree programmes. They shared a general surprise at the current level of AI capacity and they were motivated by this project to learn more about how AI might impact their own fields – by extension. Of the 20 students, only four students may have exhibited bias against AI technical writers: one English Literature, two technical writing, and one Journalism. However, the project's purpose was to gather current technologies, rather than speculate about the technologies. This distinction kept their biases abated.

AI profiles

Based on STC's TCBOK definitions, the following three AI profiles are based on the levels I, II, and III technical writers. At the end of each profile is a list of current AI writing technology capacities, along with future state AI writing technology capacities. This separation helped the Feasibility Team distinguish between what current technology can definitely do from what current technology may perhaps do under unique conditions.

Level I

A Level I AI technical writer is expected to have 0–2 years of experience, ranging from student interns to beginner technical writers. The experience can be in the technical writing field or in the field of science and technology. At entry level, the technical writer is expected to have basic writing skills and the ability to meet deadlines. They are usually supervized by others, through both mentorship and the training process. A fresh background in science and technology is a huge asset to wherever the entry-level individual may begin. An AI holding an entry-level position as a technical writer is expected to have the ability to communicate clearly and master the basics of written communication. The AI would also possess the basics of communicating with others orally or through written communication. Table 4 clusters all the current and future AI competencies separately.

Table 4: Current and future states for the level I AI Technical Writer

Current AI writing technology	Future state AI writing technology
Basic writing skills.	Complete array of writing skills.
Broad understanding of technical writing.	More detailed understanding of technical writing.
Being able to meet deadlines.	Venture into multilingual communication.
Ability to multitask assignments.	
Fluent in communication.	Taking leadership rolls in the workplace.
Mastered written communication.	Mastering the ability to create charts and graphs.
Take commands and tasks from someone of a higher position.	
Edit the work of others.	

Level II

A Level II AI technical writer is expected to have 2–4 years of experience. The AI in the second level of technical writing has mastered the basics of technical writing for each client they have to write for, adapting to different audiences. Level II writers also create detailed graphics and charts to support what is being written. The graphics and charts are built from a body of project images and a collection of scientific data respectively. At this level, the AI will also have many different firms or clients to communicate with. It is expected to communicate with each client professionally and collect requirements from client SMEs. Table 5 lists all current and future competencies.

Table 5: Itemizes current and future states for the level II AI Technical Writer

Current AI writing technology	Future state AI writing technology
▣ Basic technical writing skills.	▣ Adapting to different audiences.
▣ Ability to create graphs and charts.	▣ Ability to follow up with clients.
▣ Ability to communicate through writing to coworkers and clients.	▣ Ability to gather requirements (orally and through writing).
▣ Edit the work of others.	▣ Have a friendly but professional relationship with co-workers.

Level III

A Level III AI technical writer is expected to have 4–6 years of experience. This AI writer is fully experienced with technical writing, as well as preparing writing for translation. The AI writer could train a Level I co-worker. The AI writer may also have a detailed and massive list of clients or firms to write for. It would most likely address every co-worker by name and have the ability to recall history with clients. Not only could the AI meet deadlines at this level, it could also make sure others with less experience could potentially meet deadlines as well. This writer can take initiative in team projects. Table 6 focuses on highly experienced AI writers.

Table 6: Current and future states for the level III AI Technical Writer

Current AI writing technology	Future state AI writing technology
▣ Help the team reach deadlines.	▣ Be able to translate in a foreign language.
▣ Expert technical writers.	
▣ Train co-workers.	

Comparison of profiles against job descriptions

Three members of the Feasibility Team gathered 30 technical writer job descriptions. They identified common skills and responsibilities listed in the job ads. They considered qualifications and software proficiencies. They combined their results into a single list the rest of the Feasibility Team could match against the capabilities of the AI technical writers (levels I, II, and III). Another three members of the Feasibility Team set up three profiles that merged both technological capacity and AI technical writer level. Students referred to those profiles to gain knowledge concerning the competencies required of level I, level II, and level III technical writers.

The result of the matching activity would show how feasible it could be for the three technical writers to each enter the workforce, based on current technology and job requirements.

On Monday May 2nd 2016, 13 members of the Feasibility Team collaborated to complete the matching activity. The medium for this activity took the form of an Excel spreadsheet found on the team's shared Microsoft OneDrive folder. The purpose of the activity was to obtain a general consensus of the capabilities of the three AI writers.

Within the spreadsheet, the heading row was labeled "I, II, III, None, Lab-only" to denote the capability level of the technical writer. The option labeled "None" denoted a lack of that skill in the AI writer, and "Lab-only" indicated the primitive nature (still under testing) of the AI writer's ability to comply with the skill set. The heading column was broken up into "skills" (software) and "responsibilities." The team was directed to edit the spreadsheet by marking a letter "y" within the row corresponding to the capability of the AI writer. Figure 4 on page 333 presents a sample of the page; the numbers indicate the tally of y's. The full results are tabled in the conclusion of the Feasibility Report.

During the activity, the Feasibility Team conducted online research to gain background knowledge of the listed software in the Skills portion of the spreadsheet. Once this knowledge was acquired, the team made an educated guess on the capabilities of the AI technical writer. The team marked the rows with a letter "y" (each category should not exceed 13). Suggestions were made to mark each answer with the first letter of the student's name (to eliminate the confusion of remembering which rows one marked with "y") and questions were raised in regards to multiple markings (if level I can do it should level II and III be marked with a "y" as well?). Unfortunately, merely marking "y" was not sufficient for the

Feasibility Team; the rows had upwards of 26 votes – twice as many votes as voters. The solution was to create an 'Unclear' result to indicate when more than one column presented the majority of 13 votes.

	I	II	III	None	Lab-only		Winner	Unclear
Skills								
FrameMaker	12	9	3	1				x
Acrobat	11	2	3	1			A1	
Dreamweaver	3	11	3	1			A2	
RoboHelp	2	1	11	1	3		A3	
Microsoft Office Suite (Excel, Word, PowerPoint)	13	2	4	1			A1	
Doxygen	13	3	2	1			A1	
JavaDoc	15	1	1	2			A1	
JavaScript	1	15	8	2				x
HTML	15	1	3	2			A1	
Jira	4	15	7	1			A2	
Confluence	8	2	4	1			A1	
Linux	12	2	3	1			A1	
Windows Operating Systems	9	5	1	2			A1	
Madcap Flare	15	2	3	1			A1	
SharePoint	10	3	3	1			A1	
Access	10	3	2	1			A1	
Visio	2	9	1	1			A1	
Publisher	6	2	12	1			A3	
Endnote	5	10	4	1				x
Photoshop	2	5	12	2			A3	

Figure 4: The score sheet used during the matching activity

Conclusive factors

This Feasibility Report has reviewed the history of AI, modern AI, and provided examples of cutting edge AI. In addition, the report has showcased many examples of AI writers currently available on the market or undergoing development. The report details the profiles the Feasibility Team used to compare common technical writer job features together.

Comparison results

We divide technical writing job skills and job responsibilities into two different tables. We also include a table for the unclear comparisons. The objective is to isolate the job requirements according to the levels of technical writers.

Skills

Table 7 (overleaf) organizes software skills according to the different levels. Some of these programs are not writing programs like Office or Flare; however, an AI would still need to write in the others.

Table 7: Skills grouped by level of AI technical writer

AI level I	AI level II	AI level III
Microsoft Office Suite	JavaScript	Publisher
Acrobat	Adobe Dreamweaver	Adobe RoboHelp
Doxygen	Microsoft Visio	Adobe Photoshop
Javadoc		
HTML		
Jira Confluence		
Linux		
MadCap Flare		
SharePoint		
Microsoft Access		

Responsibilities

Table 8 focuses on the responsibilities of technical writers. These passages come from the descriptions and the bulleted job specifics.

Table 8: Responsibilities grouped by level of AI technical writer

AI level I	AI level II	AI level III
Can understand and work with basic coding concepts, terminology, and code repository systems like Git.	Can evaluate the quality of existing documentation, and make improvements, such as consolidation, to apply best practices for consistency and ease of use, which comply with documentation standards.	Responsible for the editing, rewriting, and authenticating of: technical user manuals, product articles, application papers, product descriptions, catalogues, administration guides, reference guides, brochures, commercial applications, data sheets, specification books, glossaries, framesets, publications, menus, presentations, and other materials to communicate clearly and effectively research findings, technical developments, and other news and information to a wide range of external audiences.

Table 8: Responsibilities grouped by level of AI technical writer (continued)

AI level I	AI level II	AI level III
Can deliver electronic documentation, including web content and meta tags, CBTs, online surveys and application-embedded online help.	Can collect content submitted by product managers and engineers within the organization and curate to present a consistent style, tone and voice.	Can distill complex information, documentation, and scenarios from subject matter experts, quickly and in a clear and concise style, for technical readers and readers without deep knowledge of the subject.
Can document server and appliance solutions.	Can understand and use content management tools, databases, social media (wikis and other collaboration systems, microblogs, and so on).	Possess the ability to effectively communicate with product management and software design teams.
Can analyze existing documentation for potential audit gaps.	Can plan documentation deliverables, including estimation of the size and scope of the deliverables.	Has excellent communication skills and the ability to deliver high-quality technical publications for both internal and external use.
	Can track deliverables, schedules, and resource requirements in project plan.	
	Can develop documentation within an Agile scrum environment.	
	Can develop written information regarding the development, use and support of computing systems.	
	Can work with internal teams to drive improvements.	
	Can create visual diagrams to explain processes, system architecture, etc.	
	Can identify key issues and formulate conclusions.	
	Can provide operational consulting for document/content development for internal/external clients.	

Table 8: Responsibilities grouped by level of AI technical writer (continued)

AI level I	AI level II	AI level III
	Can complete step-by-step documentations in a high volume.	

Unclear comparisons

Table 9 presents all the skills and responsibilities that did not have clear results during the Feasibility Team's matching activity.

Table 9: Unclear skills and unclear responsibilities, grouped separately

Unclear skills	Unclear responsibilities
Adobe FrameMaker	Can write policies and procedures, training materials, and job aids.
JavaScript	Can understand, interpret, and communicate complex scientific information and data quickly.
Endnote	Can analyze and interpret data to determine appropriate syntax, style, and grammatical usage.
	Can write release notes.
	Can coordinate design and writing of technical documentation.
	Can write task analyses and documentation plans to establish project goals, content, parameters, and deadlines in relation to new technical document.

Projections for the future

This Feasibility Report is a good beginning for further research. The feasibility conclusion demands a more rigorous investigation than a feasibility report. However, there is already cause for researchers in this field to look forward to when human technical writers will need to work with AI technical writers.

Future study

Even though there isn't specifically an AI technical writer, researchers need to stay vigilant. Researchers need to become involved in the direction of development as much as possible. At the same time, there is already space to consider autonomous writing. For instance, computer engineers talk about self-documenting code as the penultimate achievement of well-structured coding.

We once encountered what can be termed a Technical Writer Sweat Shop. The writers logged their writing progress in five-minute intervals. The owner said that customers want to know they are paying for writing, editing, and nothing else. Not only is that business model the first to go when AI technical writers emerge but such operations do not set the rest of the profession up for the impact of AI technical writers.

The field needs research on how current workplace writing practices can include AI technical writers on an effective writing team. Rather than resist, the field needs best practices so that human technical writers can capitalize on what will otherwise be a disruptive technology.

The Feasibility Team's matching activity is worth repeating with a team more educated in the skills and responsibilities. In addition, the method by which the matching occurs can be improved to eliminate so many unclear comparisons.

Disposition of the field

It is too late for knee-jerk reactions to AI technical writers. Whatever credibility human technical writers have gained in the workplace, they now need to consider the value they add to their career more than ever. Researchers in the field of technical writing must adopt an accepting disposition and generate research that showcases how AI technical writers increase the efficiency of human technical writers. The field has always focused on elevating writers past mere copyediting to decision-making and ownership: a member of the development team, rather than 'the writer.' With AI technical writers that provide the editorial support, human technical writers can finally prove their value in other areas.

Conclusion

The question is whether it is feasible for AI technical writers to enter the workplace. Rather than a simple answer generalized for all technical writers, this Feasibility Report splits technical writers into three groups. The answer to the feasibility question differs depending on the level.

Value of a Level I technical writer

The answer to the feasibility question is 'yes.' There are many businesses that merely need editors and do not value 'the writer'; in such cases, 'the writer' will be replaced. However, the value such businesses place on technical writers is not part of the Level I description. The kind of work such writers do can elevate what such writers typically write. In addition, the collaborative role of level I writers on a team of writers will increase. It is possible that all technical writers will become Level II writers.

Value of Levels II and III

Level II and Level III writers have many other responsibilities that require interfacing with colleagues. Some interactions seems possible with modern AI, like Nadine and Sophia; however, it is not feasible for modern AI to adopt most interactive roles. Consequently, there are a lot of collaborative roles that will magnify in importance – ultimately elevating the value of the field. In those cases, a Level I AI technical writer would do background writing, while Levels II and III dictate where the writing fits in both workplace relationships and business needs.

References

Chet (2009). Machines that Learn. *Science Musings Blog*, [Blog] 16 September. Available at: http://blog.sciencemusings.com/2009/09/machines-that-learn.html [Accessed 16 December 2016].

Giammona, B. (2009). The Future of Technical Communication: Remix. *Intercom*. May 2009. Available at: http://www.prospringstaffing.com/Resource/ GiamonaFutureofTechncial Communication.pdf [Accessed 16 December 2016].

Gollner, J. (2011). The Fusion of Business Analysis and Technical Communications. *The Content Philosopher*, [Blog]. March 11. Available at: http://www.gollner.ca/2011/03/convergence-of-analysis-and-techcoms.html [Accessed 16 December 2016].

Hanson Robotics (2016). *Sophia–Hanson Robotics*. Available at: http://www.hansonrobotics.com/robot/sophia [Accessed 16 December 2016].

Ledezma-Haight, R. (2016). X.ai Pros and How They Do It – Stephanie Ann Boyd. *X.ai Corporate Blog*, [Blog] 6 December. Available at: https://x.ai/x-ai-pros-and-how-they-do-it-stephanie-ann-boyd/ [Accessed 16 December 2016].

MIT Technology Review, (2016). How an AI alogorithm learned to write political speeches, [Blog] 19 January. Available at: https://www.technologyreview.com/s/545606/how-an-ai-algorithm-learned-to-write-political-speeches/ [Accessed 16 December 2016].

Tan, A. (2015). Human-Like Robot 'Nadine' Who Has a 'Personality, Mood and Emotions' Unveiled in Singapore. *ABC News*, 31 December. Available at: http://abcnews.go.com/Technology/human-robot-nadine-personality-mood-emotions-unveiled-singapore/story?id=36032196 [Accessed 16 December 2016].

TCBOK (2016). Technical Writer. *Technical Communication Body of Knowledge*. Available at: http://www.tcbok.org/wiki/about-tc/career-paths/technical-writer/ [Accessed 16 December 2016].

Times (2015). Did a Human or a Computer Write This? *New York Times*, 7 March. Available at: http://www.nytimes.com/interactive/2015/03/08/opinion/sunday/algorithm-human-quiz.html?_r=0 [Accessed 16 December 2016].

Warwick, K. (2012). *Artificial Intelligence: The Basics*. New York, NY: Routledge.

List of contributors

David Bird has been a technical writer for over 27 years, covering multiple technical disciplines in both the public and private sectors. He holds an Honours Degree in Information Technology, a Postgraduate Diploma in Computing for Commerce and Industry, and a National Qualifications Framework Level 7 qualification in Engineering. He is a Chartered member of the British Computer Society and is a Fellow of the Institute of Analysts and Programmers. In recent years David has also become a freelance cyber security researcher and writer. He has had many articles published in several reputable national magazines, which also have an international audience, comprising topical technical and information security subject matter.

Jody Byrne is a technical communication professional with almost 20 years' experience as a technical writer, translator, lecturer, video producer and user assistance consultant. He has written two books on technical communication and translation, and currently works for SAP Ireland where he is a member of the central video team and the head of the User Assistance Prototyping Lab.

Yvonne Cleary is a lecturer in Technical Communication and Instructional Design at the University of Limerick. Dr Cleary is Course Director for the MA in Technical Communication and E-Learning. She has published in leading technical communication journals, including *Technical Communication*, *Journal of Technical Writing and Communication*, and *IEEE Transactions on Professional Communication*.

Stephen Crabbe (Editor) is a Senior Lecturer in Applied Linguistics at the University of Portsmouth, UK. The focus of his research is technical/ professional communication and he has published widely in this area, including most recently the book *Controlling Language in Industry* (Palgrave Macmillan).

Ciaran Dodd began training in ASD-STE100 in 2002, while at Rolls-Royce, which is where she became involved in training all aspects of writing. After leaving the organization, Ciaran set up an independent consultancy specialising in interpersonal and written communication. She has extensive experience of working with major names in engineering, particularly in the defence, aerospace and automotive industries. She has taught all aspects of the English language in commercial and public organizations since 1994.

Kirstie Edwards has an MA and PhD in Technical Communication and over 25 years of experience as a freelance writer and editor in academic and commercial contexts, in the UK and Belgium. She is a writing tutor at Wrexham Glyndwr University in Wales and has published writing research and creative writing.

Marie Girard manages content strategy and architecture for IBM products. She leads unified content strategy efforts through collaboration across silos, content audits, and metrics. She is a lecturer in technical communication at Paris Diderot University, and keeps investigating how everything interrelates through the practice of yoga.

Neal Goldsmith is an experienced technical writer, content writer, copywriter, localization coordinator, researcher and software test engineer. He is currently a technical writer for dotmailer, one of the UK's leading email marketing automation platforms. He has an MA in literature and philosophy, and has performed comedy at the Edinburgh Free Fringe Festival.

Chelsea Green is a recent college graduate from Southern Connecticut State University (SCSU). She holds an undergraduate degree in English with a specialization in Professional Writing. She became interested in the field of AI research in the work place while attending SCSU. She hopes to continue this research into the future in order to fully understand the impact AI will have in the workplace within the next several years.

Andy Healey is a Senior Technical Writer at Alfresco Software, where he is the lead on end-user documentation. He recently launched their user interface writing guidelines.

Franziska Heidrich, Dr. phil., is a research assistant at the University of Hildesheim (Germany), researching and teaching in the fields of specialized translation and technical writing (German, English, French). Her primary research interests are the specialized translation process, the optimization of technical and non-technical communication, and the efficiency of technical as well as non-technical communication.

Nolwenn Kerzreho has over 10 years of experience in the technical communications industry and is the Technical Account Manager for Europe for IXIASOFT Technologies. An international speaker at leading events and academic colloquiums, Nolwenn is also in charge of developing the technical communication specialization in the European Master's in Translation at University of Rennes 2 where she introduced DITA in 2009. Nolwenn holds a Master's degree in Technical Communication, Translation, Terminology and Project Management and has international work experience in the chemical, telecom, language, and high-tech industries.

Jason Lawrence has specialized in professional communication since 2006. He has worked in global positions previous to accepting a faculty position at Southern Connecticut State University. He has published about software documentation and usability testing; he is now exploring how Artificial Intelligence professional writers can augment human efficiency.

Andrew McFarland Campbell is a technical writer with more than a decade's experience of automating documentation production for complex software products, particularly in the financial and telecommunications industries. As well as working with standard documentation tools such as DocBook and DITA he is experienced with XSLT and various programming languages. He is a Fellow of the Institute of Scientific and Technical Communicators.

Patricia Minacori is Associate Professor at the University Paris Diderot (France). A former technical translator, now Director of a department in charge of Translation and Technical Communication, head of two curricula in technical communication, she has been teaching technical and scientific translation for almost 30 years, in Graduate's and Master's degrees. Her research is on translation project management, complexity according to Edgar Morin, exchanges between technical communicators and translators, and translation assessment.

Ellis Pratt is Director and Help Strategist at Cherryleaf, a technical writing services and training company based near London, in the United Kingdom. He has over 20 years' experience working in the field of documentation, has a BA in Business Studies, and is an Associate of the Institution of Engineering and Technology. Ellis was ranked the most influential blogger on technical communication in Europe, and he is also on the management council for the Institute of Scientific and Technical Communicators.

Lorcan Ryan is a technical writer and instructional designer with over 10 years' experience in the medical, e-learning, and software industries. Lorcan has also lectured in the Global Computing & Localization postgraduate course in the University of Limerick. Lorcan's PhD won the Microsoft-sponsored LRC Best Thesis Award 2014, and he also holds an MA in Technical Communication, as well as a degree in Business Studies & Marketing.

Keith Schengili-Roberts has worked in the technical publications industry for over a quarter century, and with DITA since its launch. He now works as a DITA Specialist for IXIASOFT and as a lecturer on Information Architecture at the University of Toronto. He is the author of four technical books.

Klaus Schubert, Dr. phil., is Professor of Applied Linguistics/ International Specialized Communication at the University of Hildesheim (Germany). He was a sociolinguist at the University of Kiel (Germany), a computational linguist and project leader at the software house BSO/ Buro voor Systeemontwikkeling BV in Utrecht (Netherlands) and a consultant and technical manager at BSO/Language Technology BV in Baarn (Netherlands). For many years he served as a professor of computational linguistics and technical communication at Flensburg University of Applied Sciences in Flensburg (Germany). His main research areas are applied linguistics, translation studies, specialized communication studies, interlinguistics and applied computer linguistics.

Joe Sellman is a contract technical communicator and instructional designer who graduated from the University of Portsmouth (UK) in 2012 with an MA in Technical Communication. Joe also co-teaches a number of technical communication courses with Annette Wierstra at MacEwan University (Canada).

Mike Unwalla taught English as a foreign language to adult learners in the 1980s. In 1995, Mike became a technical writer. At an ISTC conference, he learned about ASD-STE100 and started to use it. To help him to conform to the specification, he developed a term checker for ASD-STE100.

Annette Wierstra has a MA in International and Intercultural Communications from Royal Roads University (Canada). She has been teaching technical communication and other communications courses as a sessional instructor for MacEwan University since 2008. Annette also co-owns Scriptorium Professional Writing Services (www.scriptoriumpro.com) where she works with companies on short-term contract technical communication projects.

Index